Greene Vardiman Black

A Study of the Histological Characters of the Periosteum and Peridental Membrane

Greene Vardiman Black

A Study of the Histological Characters of the Periosteum and Peridental Membrane

ISBN/EAN: 9783337396367

Printed in Europe, USA, Canada, Australia, Japan

Cover: Foto ©berggeist007 / pixelio.de

More available books at **www.hansebooks.com**

A STUDY

OF THE

HISTOLOGICAL CHARACTERS

OF THE

Periosteum and Peridental

MEMBRANE.

BY

G. V. BLACK, M.D., D.D.S.

PROFESSOR OF PATHOLOGY IN THE CHICAGO COLLEGE OF DENTAL SURGERY.

WITH 67 ORIGINAL ILLUSTRATIONS.

CHICAGO:
W. T. KEENER,
96 WASHINGTON STREET.
1887.

PREFACE.

The contents of this volume appeared in serial form in the *Dental Review*. In reviewing the matter for publication in book form, I find that the subject matter proper for a preface is included in the preliminary chapter and at other points in the progress of the work. I have concluded therefore to let it remain as written, believing that it will serve the reader fully as well, or better, than to bring it together upon this page. This volume is almost entirely a record of my own *personal* observations, written in the personal or lecture style. A division into chapters has been made for the convenience of the reader.

The volume is now offered to the profession with the hope that it may in some measure supply the want that has been felt·for a more thorough study of the histological characters of the periosteum and peridental membrane.

G. V. B.

Jacksonville, Sept. 1, 1887.

LIST OF CONTENTS.

LIST OF ILLUSTRATIONS

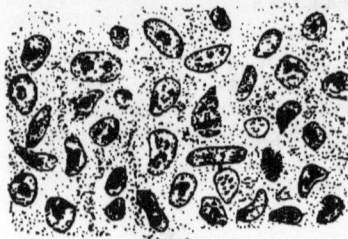

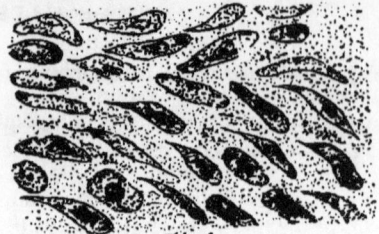

Fig. 1

Fig. 2

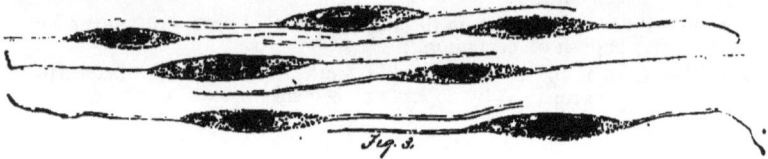

Fig. 3

Fig. 4

Fig. 5

Fig. 8

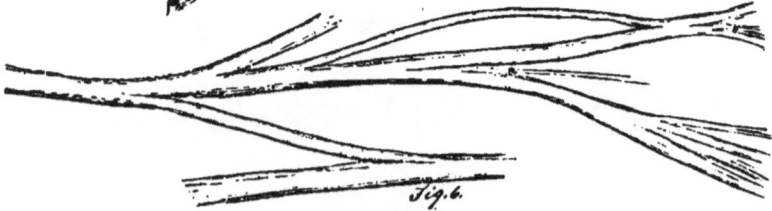

Fig. 6

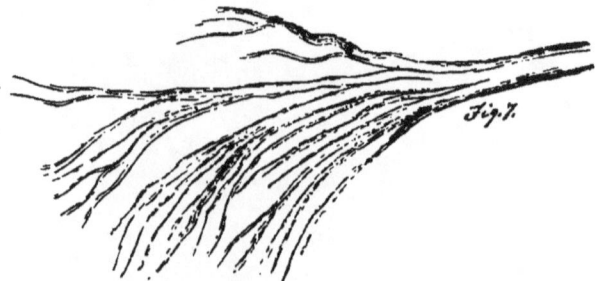

Fig. 7

Fig. 1. Embryonal connective tissue in an early stage of development, showing the cellular elements imbedded in the ground substance.

Fig. 2. The same, a little more developed, showing the cellular elements lengthening in a common direction.

Fig. 3. The cells developed in spindle forms, fibro blasts with long filaments extending from either end.

Fig. 4. The developed white fibrous tissue.

Fig. 5. Older white fibrous tissue, in which the cells are no longer seen, and showing the wave-like course of the fibers.

Fig. 6. Coarse white fibers, made up of bundles of the fine, and showing the mode of division by the splitting off of a portion of the fibers of the bundle.

Fig. 7. Coarse fiber breaking up into fine fibers.

Fig. 8. Cross sections of coarse fibers showing some of their various forms.

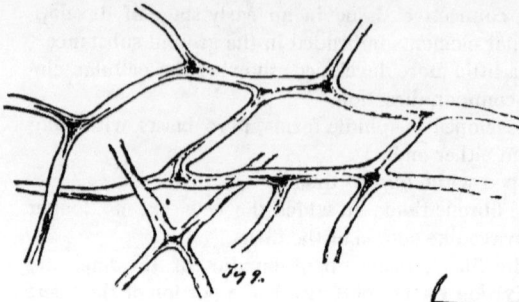

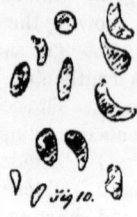

Fig. 9.

Fig. 10.

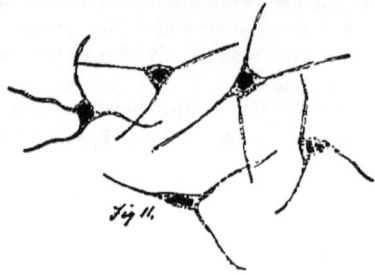

Fig. 11.

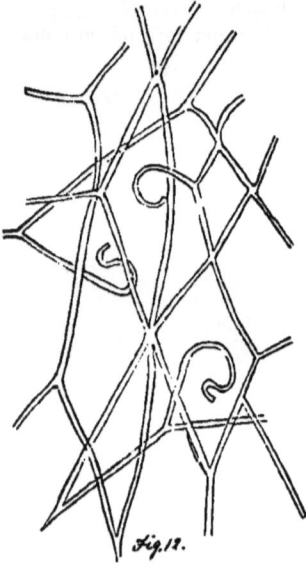

Fig. 12.

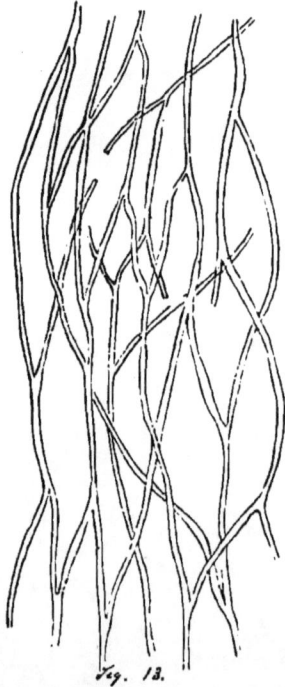

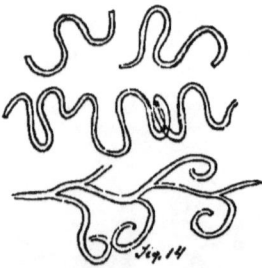

Fig. 14.

Fig. 13.

Fig. 15.

Fig. 16.

Fig. 9.   Reticular fibers, showing the mode of division and the multi-polar, or irregular star forms of the cells at the divisions.

Fig. 10.   Cross sections of the reticular fibers, showing some of their forms.

Fig. 11.   Connective tissue cells from which reticular fibers are developed.

Fig. 12.   Network of elastic fibers from the point of reflection of the mucus membrane of the lip from the gums.

Fig. 13.   Network of elastic fibers teased out from elastic tendon, and showing the usual mode of division.

Fig. 14.   Elastic fibers, showing their disposition to curl up when cut or broken.

Fig. 15.   Cross sections of elastic fibers, showing their forms as seen in a group passing between coarse white fibers.

Fig. 16.   Tissue of the dental pulp, in which the development of the cells is not followed by any considerable formation of fibers.

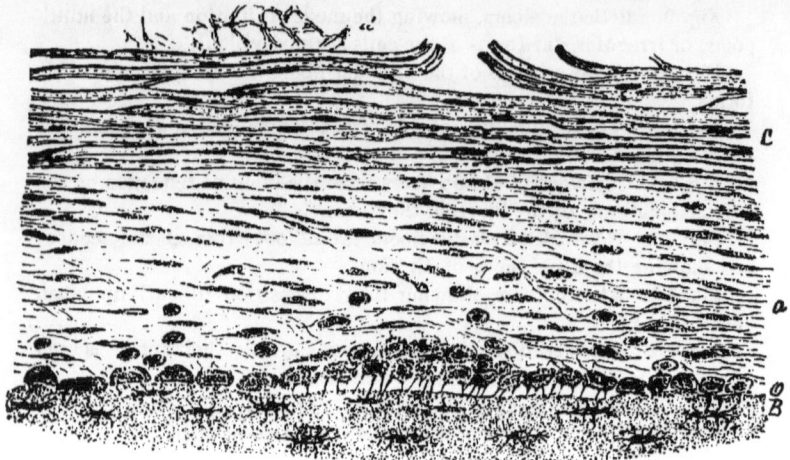

Fig 17. 12th in immersion Obj.

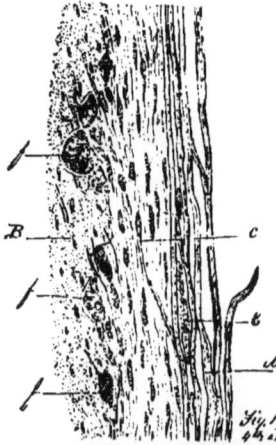

Fig 19.
4th in Obj.

Fig 18. 12th in im. Obj.

Fig. 17. Non-attached periosteum from the shaft of the femur of the kitten. *B*. Bone. *O*. Layer of osteoblasts. In the central portion of the figure they have been pulled slightly away from the bone, displaying the processes to advantage. It will be observed that the fibers of the periosteum do not enter the bone. *a*. Inner layer of fine white fibrous tissue (osteogenetic layer) showing the nuclei of the fibroblasts and a number of developing connecting tissue cells, which probably become osteoblasts. *c*. Outer layer, or coarse fibrous layer, in which fusiform fibroblasts are also rendered apparent by double staining with hematoxylin and carmine. *d*. Some remains of the reticular tissue connecting the superimposed tissue with the periosteum.

Fig 18. Periosteum from the shaft of the tibia of the pig, lengthwise section, showing the complex arrangement of fibers in the coarse, or outer fibrous layer that sometimes occurs under muscles that perform sliding movements upon it *B*. Bone. *O* Layer of osteoblasts The tissue has been pulled slightly away from the bone in mounting the section, and part of the osteoblasts have clung to the bone, some have clung to the tissues, while others are suspended midway, their processes clinging to each. *a*. Layer of fine fibers. Inner or osteogenetic layer of the periosteum. *b*. First lamella of the coarse or outer fibrous layer, the fibers of which are, in this case, circumferential, exposing the cut ends. It will be observed that there are ten lamellæ in the make-up of the outer layer, the lengthwise and circumferential fibers alternating. The ones marked *f*, and *i*, are very delicate ribbon-like forms, which have shifted from their normal position in the mounting of the section, so as to present their sides to view instead of their ends, thus displaying their structure to advantage. The illustration shows how readily separable these lamellæ are. *l*. Reticular tissue.

Fig. 19. Periosteum from the lower end of the femur of the kitten at a point where the enlarged end next the joint is being trimmed down for the elongation of the shaft, showing the fibers of the periosteum included in, or entering the bone, forming its attachment, the absence of osteoblasts and the presence of osteoclasts by which the outer portions of the bone are being removed. *B*. Bone. *c*. Osteogenetic, or inner layer of periosteum. *d*. Outer layer, a part of which seems to have been torn away *E*. A few circumferential fibers. *f.f.f*. Osteoblasts lying in the lacunæ of Howship, or excavations in the bone made by these cells.

Fig. 20. Attached periosteum from beneath the attachment of the muscles of the lower lip of the sheep. *a*. Bone. *B*. Osteoblasts, with the fibers emerging from the bone between them. *c*. Inner layer with fibers decussating and joining the inner side of the coarse fibrous layer in opposite directions. This is rather an unusual form of this layer of the periosteum. *D*. Coarse, fibrous layer. *E*. Attachment of muscular fibers.

Fig. 21. 4·4 in. Obj.

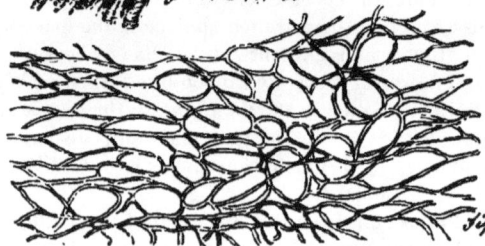

Fig. 22. 12ᵗᵉⁿ im Obj.

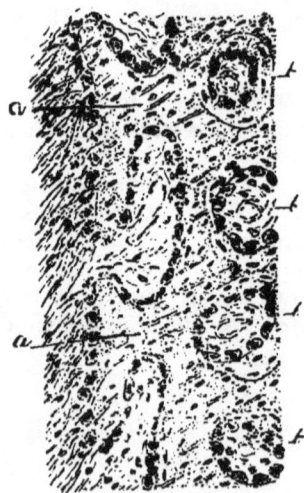

Fig. 23. 4ᵗᵉ in. Obj.

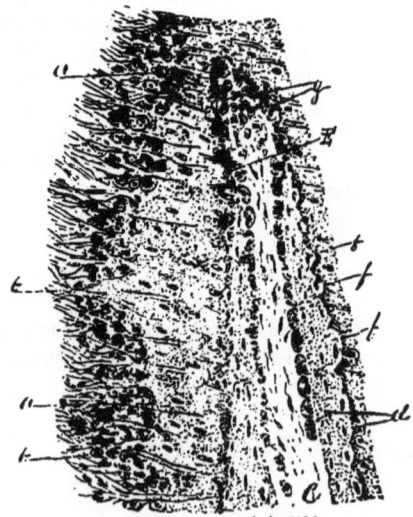

Fig. 24. 4·4 in. Obj.

Fig. 21. The more usual form of the attached periosteum. *A*. Bone, showing the residual fibers (penetrating fibers of Sharpey) within its substance and passing out between the osteoblasts *B*, and breaking up into fine fibers, which form the internal layer of the periosteum. These are also seen protuding from the broken margins of the section at *g. g. g*. *D*. Blood-vessels which are cut across. They occur mostly in the inner layer, very close to the under side of the outer layer. A number of them are seen. *H*. Small nerve bundles. *F*. Attachment of muscular fibers. It will be noted that the Haversian canals at *h. h. h. h*., and at other points, are filling up with bone that has no residual fibers.

Fig. 22, Network of elastic fibers from the coarse fibrous layer from a section of the same series, as Fig. 21, after dissolving out the coarse fibers with caustic potash. High power.

Fig. 23. Bone, with portion of inner layer of attached periosteum, and penetrating fibers. The section is cut across the Haversian canals, and it shows the manner of the formation of these in the surface of the growing bone at *a. a*. by the upward growth of spiculæ of bone which then spread out and join with others, thus bridging over and forming canals. At *b. b. b. b*. four Haversian canals are seen lined with osteoblasts. Around each of these, fresh bone is being deposited, which may be recognized by a slight difference in shade, but especially by the fact that the bone corpuscles lie in a different position from others in their neighborhood, and the fact that this bone has no residual fibers. It should be noted that this formation of canals immensely increases the area upon which osteoblasts may build.

Fig. 24. Bone, with a more solid growth of surface, and with osteoblasts much crowded between the fibers of the periosteum as ther emerge from the bone. Only a part of the inner layer of periosteum is shown. *a. a*. Osteoblasts several layers deep between the fibers of the periosteum. *b. b*. Spiculæ of bone growing up into the periosteum, apparently following the line of a particular fiber. *C*. A Haversian canal that seems to have been excavated in the bone, and is being filled by deposit of new bone on its walls. This new deposit of bone is distinguished by a somewhat lighter shade, and the difference in the direction of the long axis of the bone corpuscles, and the absence of residual fibers. Osteoblasts appear in this portion of the canal. The margins of the secondary formation, show the bay-like forms usual in the absorption of bone. Above the line drawn at *E*, no secondary bone is found, and osteoclasts, *g. g*, are seen instead of osteoblasts. In this portion the excavation is going on. In this way the bone, with residual fibers, is removed and bone deposited in which these do not appear.

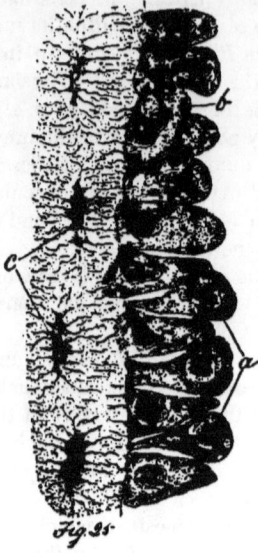

Fig. 25.

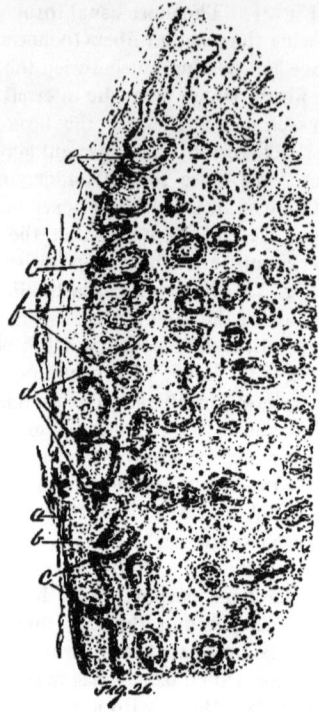

Fig. 26.

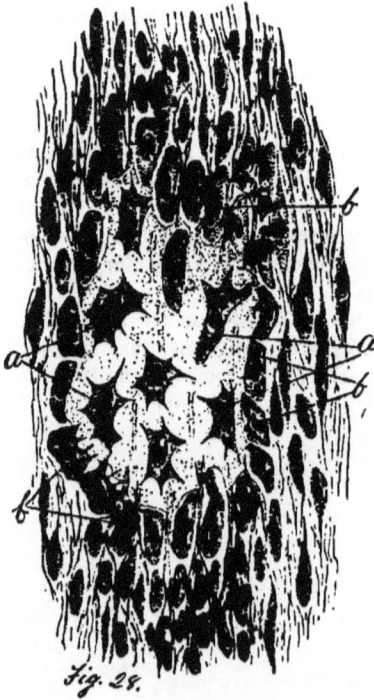

Fig. 28.

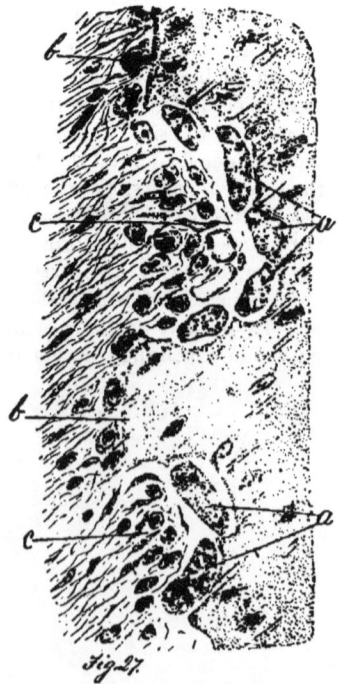

Fig. 27.

Fig. 25, 12th in. immersion obj. Higher eye piece. Margin of growing bone upon which the osteoblasts are very much crowded. *a*, Osteoblasts reaching to the surface of the bone by extending process-like prolongations. *b*, A cell that seems to be flattening down upon the surface of the bone. *c*, Bone corpuscles, the processes of which are seen radiating in the bone matrix. Processes are also seen extending into the bone from some of the osteoblasts.

Fig. 26, 1-2 in. obj. Cross section of a young growing bone, showing the Haversian canals and the plan of their subperiosteal formation. *a*, Outer layer of periosteum. *b*, Inner layer of periosteum. *c, c*, Spiculæ of bone growing outwards into the tissue of the inner layer of periosteum. *d*, Other and older spiculæ spreading out at their summits, forming portions of arches. *e*, Other spiculæ, the arches of which are about closing to form Haversian canals. *f*, Complete Haversian canals, many of which are seen in the illustration.

Fig. 27, 1-8 in. obj. Absorption of bone under attached periostéum. *a, a*, Osteoclasts lying in deep excavations in the surface of the bone. *b, b*, Surface of bone, showing the fibers of the periosteum implanted in it. Residual fibers appear in the bone. It will be noted that these fibers are removed with the bone by the absorptive process. *c, c*, Masses of fetal tissue filling the areas formed by the absorption.

Fig. 28, 12th inch immersion obj. Intra-membranous formation of bone. An island of bony deposit. *a, a*, Bone corpuscles. *b, b*, Osteoblasts. It will be seen that these lie between the fibers of the membrane, so that in certain positions the osteoblasts lie with their ends to the forming bone. And for the most part the long axes of the bone corpuscles have a similar direction.

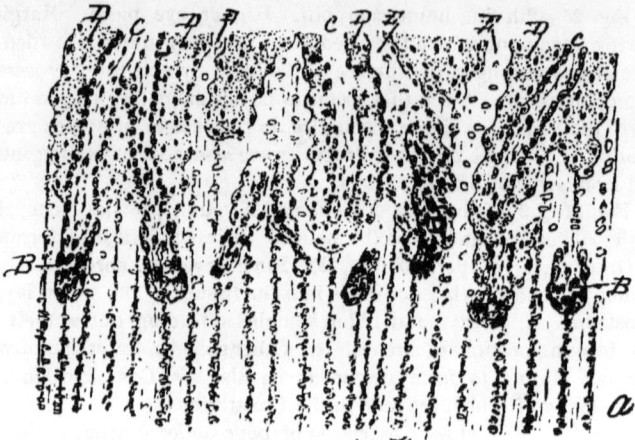

Fig. 29.

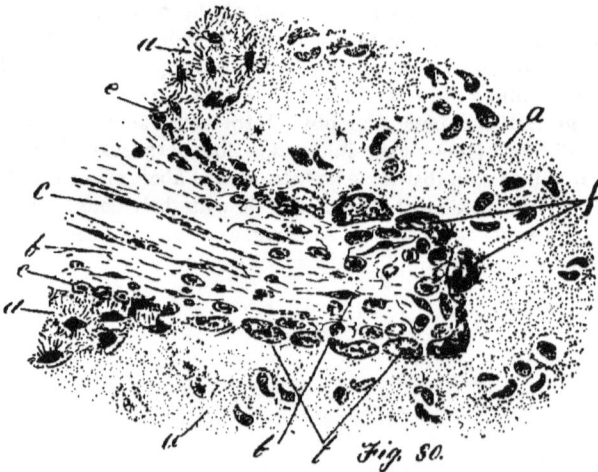

Fig. 30.

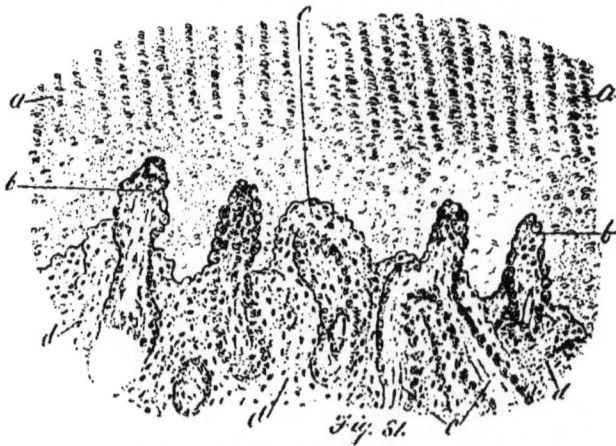

Fig. 31.

Fig. 29, 1-4 in. obj. Growth of bone under the attachment of the Tendo Achillis in a young lamb. *A*, Fibers of tendon partially converted into fibro-cartilage. The cartilage cells are seen mostly between the tendon fibers. *B, B,* and *c, c, c,* Canals advancing from the bone beneath into the tendon. *D, D, D,* Bone deposited upon the walls of the canals forming Haversian systems laid upon, or among the tendon fibers. *E,* Portions of the tendon fibers still remaining deep among the Haversian systems of bone.

Fig. 30, 12th in. immersion obj.—reduced. *A* Single canal as shown at *b,* fig. 31, very much enlarged. *a, a,* Cartilage. *b, b,* Tissue of canal. *c,* Blood vessel. *d, d,* Bone. *e, e,* Osteoblasts. *f, f,* Chondroclasts. In both these figures the bay-like excavations of the absorption cells are seen in the canals, and at the margins of the bone deposited in these.

Fig. 31, 1-4 in. obj. Epiphysal intra-cartilaginous formation of bone from head of tibia of young lamb. *a, a,* Cartilage, the cells of which have fallen into rows, but have become scattered between the letters *a,* and *b, b.* *b, b,* Haversian canals advanced from the bone into the cartilage. It should be noticed that these are lined with chondroclasts where the absorption of cartilage is in progress, and with osteoblasts when bone is being deposited. *C,* Blood vessels. *d, d, d.* Bone, which is extended into the cartilage by the filling of the canals formed by absorption as shown at *e.*

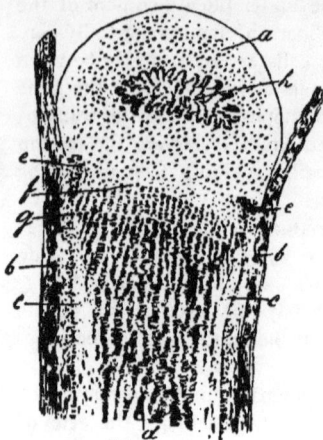

Fig. 32.

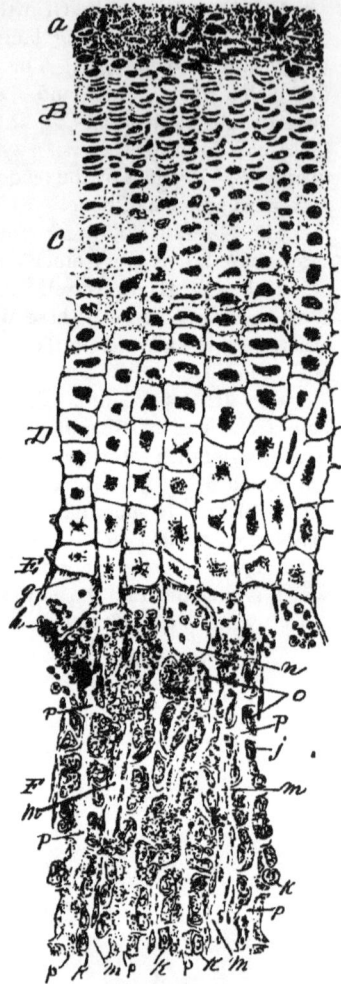

Fig. 33.

Fig. 34.

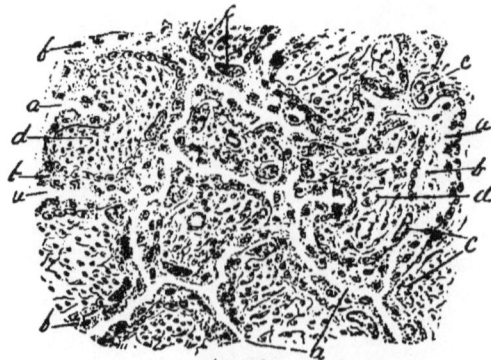

Fig. 35.

Fig. 32. Low power. Central section of the head, and portion of the shaft of the tibia from young kitten, showing diaphysal intra-cartilaginous formation of the bone at *d*, and the beginning of the epiphysal at *h*. *a*, Cartilaginous head of bone. *b, b,* Periosteum. *c, c*, Layer of subperiosteal bone. *e*. Periosteal notch ; the point to which the subperiosteal formation of bone extends. *f*, Beginning of change in the cartilage cells where they form rows. *g*, Line of absorption of the cartilage. At *d*, the darkened portion reaching up to the line *g*, shows the portion occupied by the bone marrow, and the light portions the bone formed.

Fig. 33, 1-8 in. obj. The changes which occur in diaphysal intracartilaginous formation of bone. *a*, Cartilage unchanged. At *B*, the cells have become smaller and have fallen into rows. At *C*, the cells are enlarged in their short diameters, or in the direction of the length of the shaft of the bone. At *D*, the growth of the cells has reached its limit. The matrix begins to calcify. At *E*, the capsules of the cells are opened by the advance of the absorbent tissue. F, Area of the formation of bone. *g*. Apparently some glutinous remains of the cell body clinging to the walls of the capsule. *h*, Small, round cells—marrow cells. *p, p, p*, Remains of the cartilage matrix. *j*, Osteoblasts applied to the remains of cartilage matrix, but no bone is seen. *K, K, K*, Osteoblasts and a layer of bone deposited on the remains of cartilage matxir. *m, m, m, m*, Blood vessels. *n*, Capsule which seems to have been just opened and the marrow cells seen in the act of crowding into it. *o*, Fusiform cells. Many of these appear in this portion of the figure, and seem peculiar to this location.

Fig. 34. Supplement to fig. 33, taken from another portion of the section and showing the marrow cells applied closely to the walls of the capsules next to be opened. *a*, Cartilage. *b*, Fusiform cells filling closely the last capsule opened in that row. *c. c*, Round, marrow cells filling other capsules in the same manner. *d*, Unabsorbed remains of cartilage matrix.

Fig. 35, 1-8 in. obj. From a cross section of a rib of a young kitten at a little distance (boneward) from the change from cartilage to bone; showing the large Haversian canals with the remains of the cartilage matrix enveloped in the bone formed. *a, a, a, a*, Remains of cartilage matrix which, in the figures, is left white. *b, b, b, b*, Bone deposited on remains of cartilage matrix, and generally covered with osteoblasts, but at *c, c, c, c*, and other points, osteoclasts are quite plentifully distributed. While in one part bone is being deposited, in another it is being removed, and in the end all the cartilage matrix disappears.

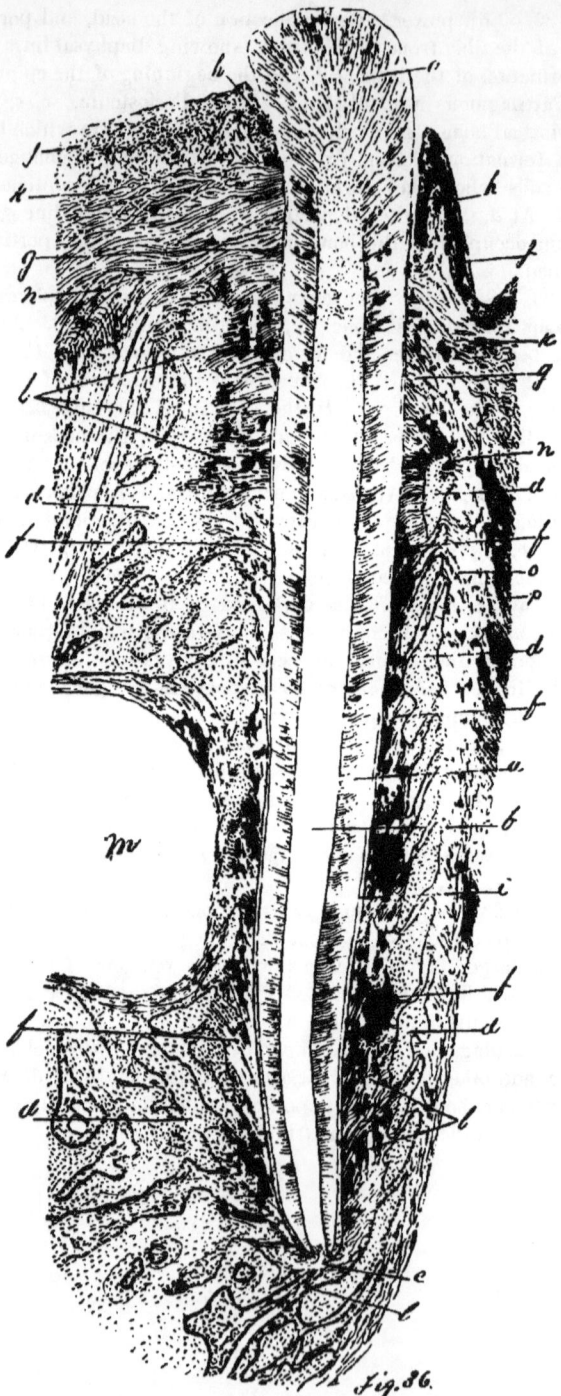

Fig. 86.

Fig. 36, 2 in. obj. Lengthwise section of small incisor tooth of kitten with its membrane and alveolus. The portion included in the illustration is one-fourth in. long. *a, a,* Crown of tooth and dentine. *b,* Pulp chamber and root canal. *c,* Cementum. *d, d, d, d,* Alveolar walls. *e,* Apical space and apical foramen. *f, f, f, f,* Body of peridental membrane, showing particularly the arrangement of its principal fibers, their direction, etc. *g, g,* The cervical portion of the peridental membrane, showing the relation of its fibers to the gingivus *h,* the tangled mass of fibers forming the gums *k,* and the periosteum *n, n,* of the outer surface of alveolar wall. *h, h,* Gingivus. *j, j,* Epithelium. *k, k,* Coarse fibrous tissue of the gums. *l, l, l,* Bloodvessels traversing the peridental membrane. A section showing the smallest number of these was selected, for the reason that the fibrous arrangement is less distorted. *m,* Saculus of permanent tooth. The fibers of the peridental membrane become continuous with those of the periosteum at *n, n. o,* Periosteum. *p,* Attachment of labial muscles. The intention of the illustration is to give a full view of the arrangement of the fibers of the peridental membrane, and the relations of the tooth, membrane, and alveolar wall.

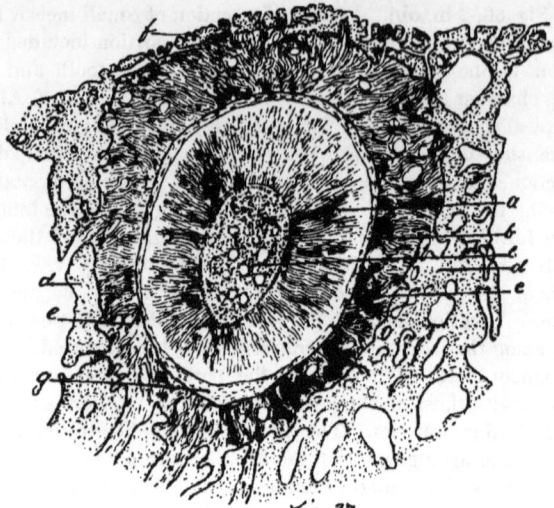

Fig. 37.

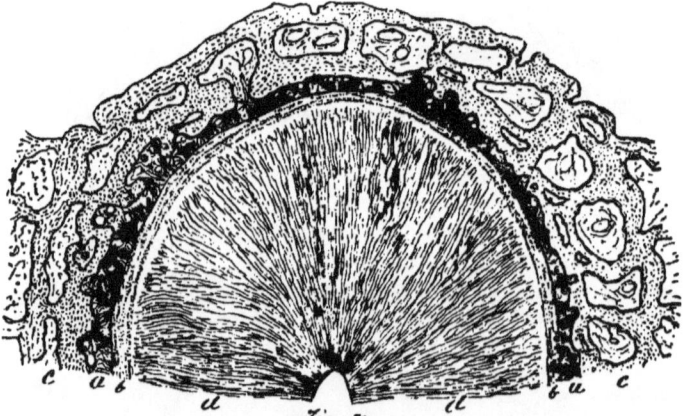

Fig. 38.

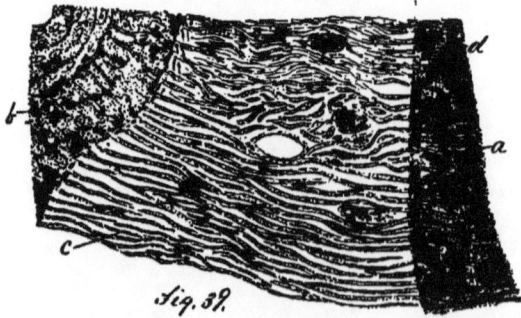

Fig. 39.

Fig. 37, 2 in. obj. Cross section of the root of a temporary incisor with the peridental membrane and alveolar walls, at about the middle of the lower third of body of the peridental membrane, showing the direction of the fibers of the membrane, and the position of the blood-vessels. *a*, The dentine. *b*, Cementum. *c*, Pulp. Its blood-vessels are shown. *d*, *d*, Alveolar wall, septi between the teeth. *e*, *e*, Peridental membrane. The direction and arrangement of its fibers have been carefully represented; also the position and relative size of its blood-vessels. *f*, Thin portion of the anterior alveolar wall. *g*, Hypertrophy of the cementum.

Fig. 38, 2 in. obj. Cross section of cuspid tooth with peridental membrane and alveolar wall cut through the thickened rim at the gingival portion of the alveolar wall, from a man forty years old. The membrane was very thin and firm, and a large piece of the anterior wall of the alveolus adhered to the tooth when extracted. It therefore represents an extremely thin membrane, while fig. 37 represents one that may be regarded as thick. *a*, *a*, Peridental membrane. *b*, *b*, Cementum. *c*, *c*, Alveolar process. *d*, *d*, Dentine. It will be observed that most of the blood-vessels of the peridental membrane lie in depressions in the alveolar wall.

Fig. 39, $\frac{1}{4}$ in. obj. Fibers of the peridental membrane passing from the cementum *a*, to the alveolar wall *b*. The section is from the root of a first molar of a man about seventy years old. The point chosen for this illustration includes a portion of a strong band of solid fibers *c*, which pass unbroken from the cementum to the bone. More generally, the fibers, after emerging from the cementum, break up into finer fibers or fasciculi, as at *d*. This form of the fibers is better shown in fig. 42.

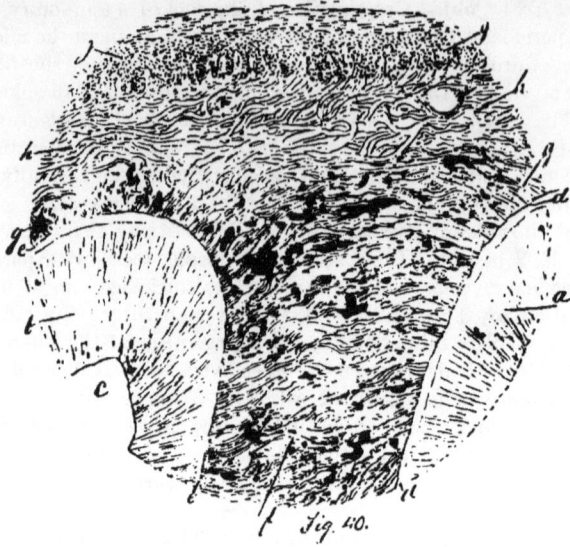

Fig. 40.

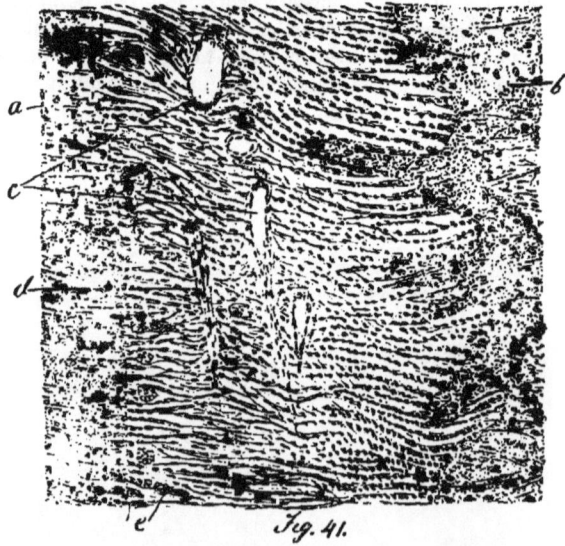

Fig. 41.

Fig. 42.

Fig. 43.

Fig. 40, 2. in obj.   Cross section of the central and lateral incisors below (toward the crowns) the rim of the alveolar wall, or through the necks of the teeth, showing the tissue of the septum and of the gums anteriorily.   *a,* Portion of central incisor.   *b,* Lateral incisor.   *c,* Pulp chamber of lateral incisor.   *d, d,* Cementum of central incisor.   *e, e,* Cementum of lateral.   *f,* Fibers of the peridental membrane extending from tooth to tooth continuously.   These are fixed in the cementum of each tooth, and form the tissue of the septum.   *g, g,* Fibers of the peridental membrane, which join with the coarse fibrous tissues of the gums *h, h.*   *j, j,* Epithelial covering of the gums.

Fig. 41, ½ in. obj.   Peridental membrane from perpendicular section of a tooth of the pig, stained with nucleus tinting dye.   *a,* Cementum. *b,* Bone.   *c,* Blood-vessels cut diagonally.   *d,* Nerve bundle.   *e,* Lymphatics.   A number of these are seen near the cementum.   The principal fibers are transparent, while the interfibrous tissue is stained.   The cellular elements appear in rows between the principal fibers, which are large and strong near the bone, and only partially break up into fasciculi in the central part of their length.

Fig. 42, 12th in. obj. (reduced.)   Fibers emerging from the cementum and breaking up into fasciculi.   From the peridental membrane of a molar of an aged person.   This represents the more usual form of the principal fibers, as seen in old age in man.   They pursue a somewhat wavy course, and generally the identity of the individual fiber is lost. They are inserted into the bone in compact bundles similar to those of the cementum.

Fig. 43, 12th in. obj. (reduced.)   A group of fibers emerging from the cementum and radiating fan-like.   On either side, the principal fibers are absent for a little space, which is filled with indifferent tissue. From the apical space (at the apex of the root) of a bicuspid of an old person.

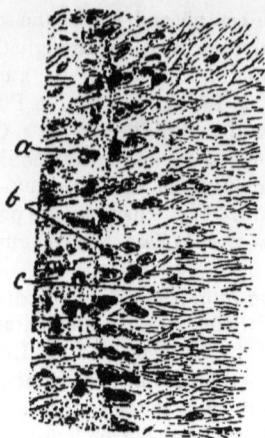

Fig. 45.

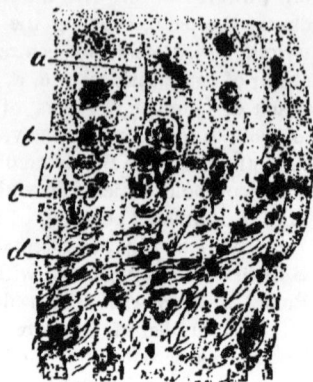

Fig. 46.

Fig. 47.

Fig. 48

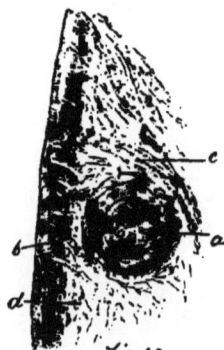

Fig. 49.

Fig. 45, 12th in. obj. (reduced.) From section including a portion of the alveolar wall, and portions of the peridental membrane, showing the osteoblasts. *a*, Bone. Inner margin of alveolar wall, showing residual fibers. *b*, Osteoblasts. Developing cells are seen in the neighborhood. *c*, Fibers of the peridental membrane. It will be noted that these spring from the bone as solid fibers and immediately break up into fasciculi.

Fig. 46, 12th in. obj. (reduced.) From section including a portion of the alveolar wall, and fibers of the peridental membrane at a point where these latter are large and compact, and with interfibrous tissue between them. *a*, Bone, showing the large residual fibers. *b*, Osteoblasts filling spaces between the fibers. *c*, Principal fibers of peridental membrane, which at this point maintain the solid form far out from the bone. *d*, Interfibrous tissue consisting of fibroblasts and fibers which lie between the principal fibers and pursue an independent course. Compare with fig 45.

Fig. 47, 12th in. obj. (reduced.) A lymph follicle or node from near the gingival border of the peridental membrane. *a, a, a,* Lymph-cells seemingly inclosed within enlarged lymph-ducts. *b, b,* Capillary vessel.

Fig. 48, ⅓ in. obj. Lymp-ducts crowded with lymphoid cells. From a section taken horizontal to the surface of the cementum, but a very slight distance from it. Cross cuts of these are seen at *c, c,* in fig. 50.

Fig. 49, ⅓ in. obj. Calcospherite-like spherule in the tissues of the peridental membrane. *a*, Spherule. *b*, Cementum, showing the fibers of the peridental membrane springing from it. *c*, Principal fibers of membrane. *d*, Indifferent tissue. For a small space no fibers are attached to the cementum.

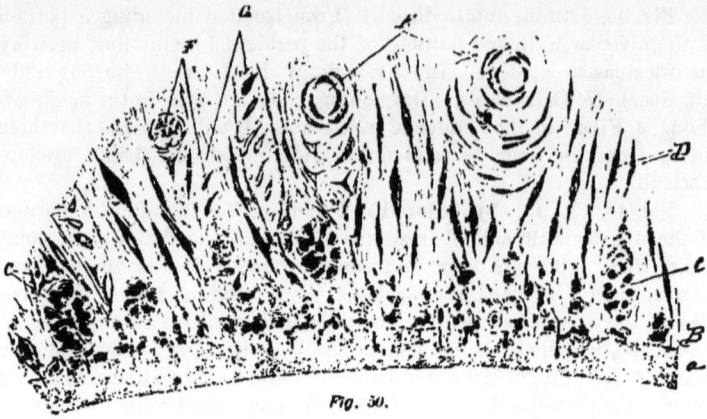

Fig. 50.

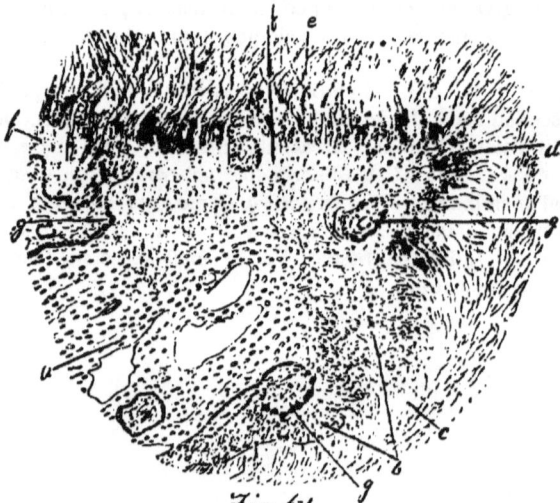

Fig. 51.

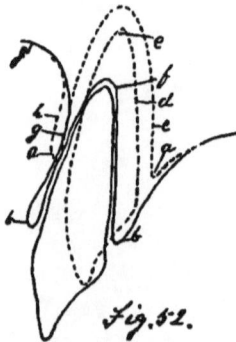

Fig. 52.

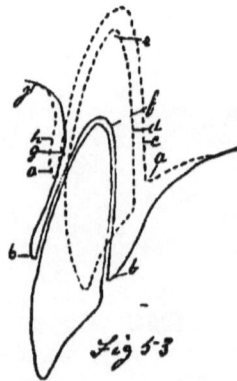

Fig. 53.

Fig. 50, 12th in. obj. (reduced.) Cementum and portion of the peridental membrane from the sheep. From a cross section of the tooth. *a*, Cementum. B, Cementoblast lying between the fibers, which latter break up into fasciculi immediately after leaving the cementun. *c, c,* Cross section of the lymph follicles or nodes. D, Fibroblasts. E, Blood vessels. These are accompanied by a large amount of interfibrous, or indifferent connective tissue. F, Nerve bundle. G, Fasciculi of fibers pursuing a direction different from the main trend of the principal fibers.

Fig. 51, ½ in. obj. Rim of the alveolar wall, from a perpendicular section. *a*, Haversian bone, which is left without stippling to render it more apparent. *b*, Subperiosteal bone, showing residual fibers. *c*, Periosteum. *d*, Extreme gingival margin of the alveolar wall. *e*, Fibers of the peridental membrane. *f*, Bone formed by the osteoblasts of the peridental membrane. *g, g, g,* Points at which the absorption of bone is in progress.

Figs. 52 and 53. Diagramatic illustration of the movement of a central incisor during the growth of the alveolar process between the age of twelve and twenty-one years. The broken lines represent the tooth and its alveolus at twelve years of age, and the solid the same tooth at twenty-one.

Fig. 52 represents the minimum movement, while fig. 53 represents the maximum movement, as ordinarily observed. The figures are lettered alike. The growth of the process is represented by the movement from *a, a* to *b, b.* The tooth is carried forward with this growth, and the alveolus is filled with new bone from the line *e* to the line *f*.

Fig. 54.

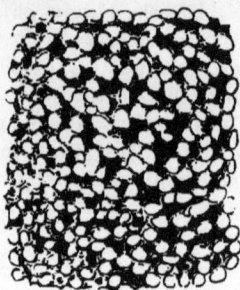
Fig. 55.

Fig. 57.

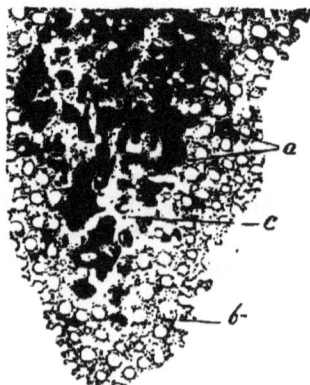
Fig. 56.

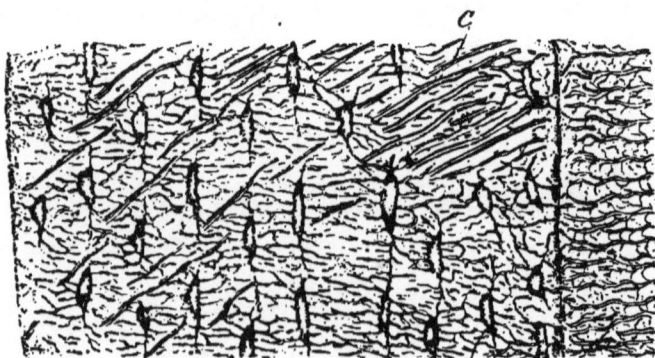
Fig. 58.

Fig. 54, 12th in. obj. Cementoblasts isolated to show the peculiar irregular forms of these cells.

Fig. 55, 12th in. obj. Cementoblasts *in situ*, with cross sections of the principal fibers of the peridental membrane of the pig, from a section cut horizontal to the surface of the cementum and including these cells. It will be seen that the cementoblasts fill all the space not occupied by the principal fibers. (In figure 57 c, the cementoblasts are seen as they appear in section perpendicular to the surface of the cementum.)

Fig. 56, 12th in. obj. Section of cementum of pig cut horizontal to and near the surface of the root of the tooth, showing cross sections of the included fibers. b, Thin margin of section, from which the fibers have fallen out of their alveoli. c, A little thicker portion in which the fibers remain. It will be noticed that from shrinkage the fiber is a little small for its alveolus, so that it is slightly separated from one side. a, Cement corpuscles

Fig. 57, ⅓ in. obj. Perpendicular section of the cementum of a pig, showing the included fibers of the peridental membrane. e, Margin of cementum showing fibers passing from the cementum to the peridental membrane, and the layer of cementoblasts with other cells in the neighborhood. f, Lymphatics. d, d, Fibers protruding from broken margin of section. a, Dentine. b, Junction of dentine and cementum.

Fig. 58, 12th in. obj. Cementum of pig from the dried section. a, Dentine. b, Lacunæ of cementum with canals anastomosing with each other. c, Imperfectly calcified fibers. It will be noticed that a few of the dentinal tubes pass through into the cementum.

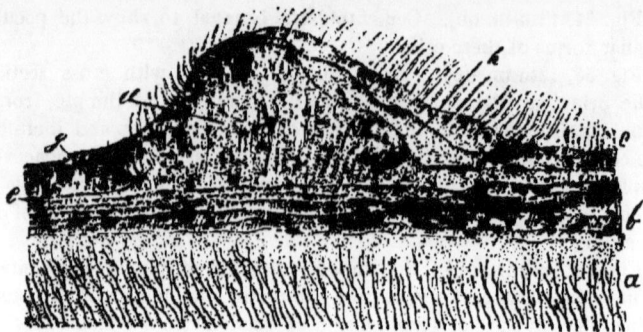

Fig. 59.

Fig. 60.

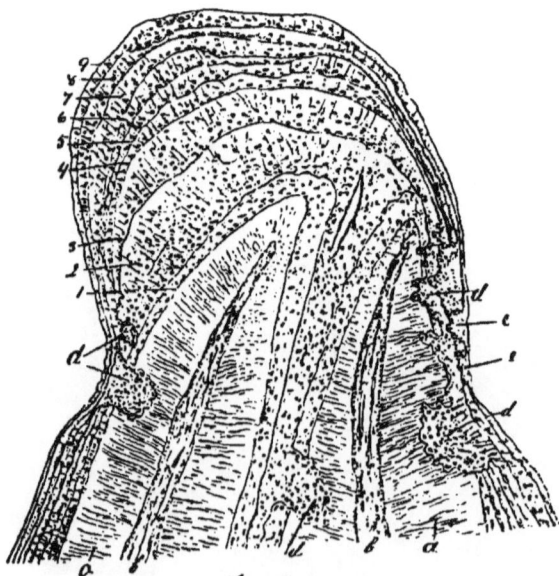

Fig. 61.

Fig. 59, 1 in. obj. Hypertrophy of the cementum on the side of the root of a lower molar near the neck of the tooth. From a lengthwise section, man. *a*, Dentine. *b*, Cementum. *c*, Fibers of peridental membrane. From *b* to *c* the cementum is normal, and the incremental lines fairly regular, but at *d*, one of the lamellæ is greatly thickened. At *e*, this lamella is seen to be about equal in thickness with the others.

The next two lamellæ are thin over the greatest prominence, but one is much thickened at *g*, and both at *h*. These latter seem to partially fill the valleys which were occasioned by the first irregular growth.

Fig. 60. 1. in obj. Hypertrophy from root of cuspid. man, in which the irregularity is confined to the first lamella. *a*, Dentine. *b*, Thickened first lamella. *c*, Subsequent lamellæ, which are seen to be fairly regular.

Fig. 61, 2 in. obj. Apex of root of an upper bicuspid tooth with irregularly developed cementum. *a, a*, Dentine. *b, b*, Pulp canals. The lamellæ of cementum are marked 1, 2, 3, etc. *d, d, d*, Absorption areas that have been refilled with cementum.

It will be seen that the apices of the roots were originally separate, but became fused with the deposit of the second lamella of cementum, and that in this the irregular growth began and was most pronounced. It has continued through the subsequent lamellæ, but in less degree. It will also be noticed that the absorption areas, *d, d, d*, have proceeded from certain lamellæ. That between the roots has broken through the first lamella and penetrated the dentine, and has been filled with the deposit of a second lamella. Other of the absorptions have proceeded from lamellæ, which can be readily made out. The small points, *e*, seem to have been filled with the deposit of the last layer of the cementum, while others have one, two or more layers covering them.

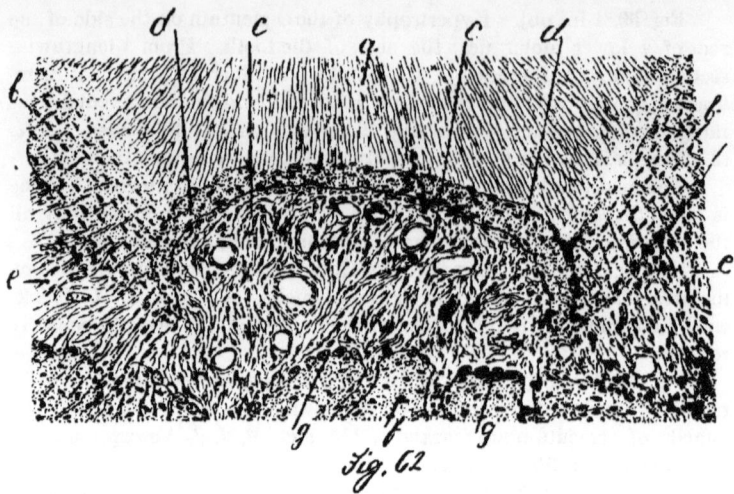

Fig. 62

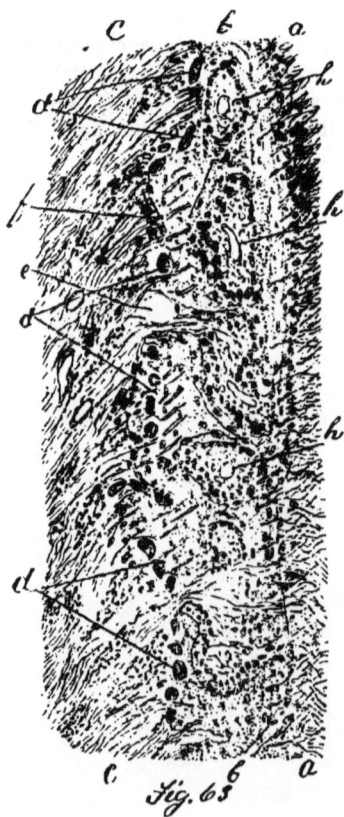

Fig. 63

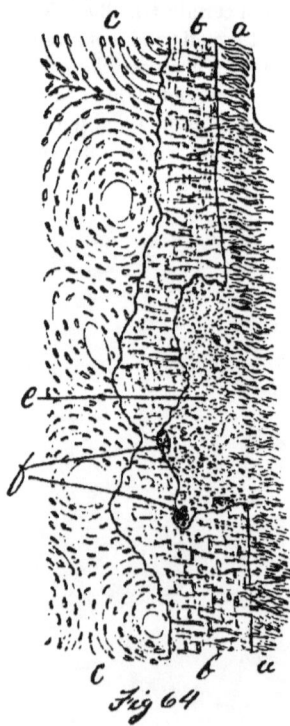

Fig 64

Fig. 62, ¼ in. obj. Cross section of the root of a temporary incisor tooth of the pig, showing a large area of absorption which is partly filled in with cementum.

a, Dentine. b, b, Cementum. c, c, Area of absorption. It will be noticed that in this area all of the cementum and a considerable portion of the dentine has been removed. d, d, Cementum that has been laid down upon the surface of the dentine and cementum alike. e, e, Peridental membrane. f, Portion of bone forming the wall of the alveolus that has grown forward into the area of absorption. g, g, Osteoclasts which are removing these bony projections. The bone which has been advanced here to take the place of the absorbed area is being removed again in compliance with the rebuilding of the cementum, which is in progress.

Fig. 63, ½ in. obj. Portion of the anterior alveolar wall of an incisor that is being absorbed. a, a, Portion of the inner layer of the periosteum. b, b, Bone forming a portion of the anterior wall of the alveolus. It will be observed that it contains a number of Haversian canals, h, h. c, c, A portion of the peridental membrane. d, d, d, Osteoclasts which are in the act of removing the bone, thus widening the alveolus. e, Space from which a large osteoclast has probably fallen during the preparations of the section. It will be noticed that where the osteoclasts are removing the bone, the fibers of the peridental membrane are detached and some little space is occupied by tissue of a fetal type, but in the spaces between the groups of osteoclasts the fibers are firmly attached to the bone. At f, there seems to be a little new bone formed to which fibers are attached. In this way bone seems to be removed, part by part, and the attachment of the membrane maintained.

Fig. 64, ¼ in. obj. Portion of the alveolar wall of a cuspid tooth of an old person, showing absorptions a, a, Portion of the peridental membrane. b, b, Portion of bone that seems to have been built on to supply an area of previous absorption. e, A recent absorption area. At f, three osteoclasts are seen. It will be noted that the fibers of the peridental membrane are detached throughout this area of absorption and the space is occupied by tissue of a fetal type. It should also be noted that the Haversian systems of the bone had been cut into by the previous absorption, removing portions of the rings of the Haversian systems. Residual fibers are seen in the bone b, but there are none in the Haversian bone c.

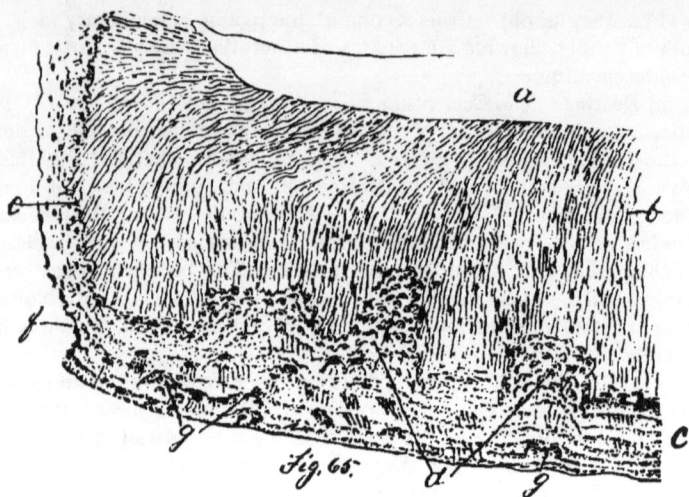

Fig. 65.

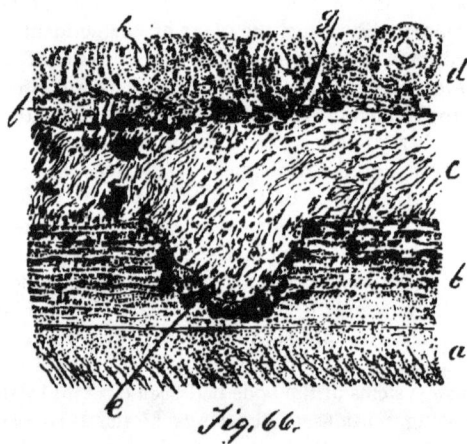

Fig. 66.

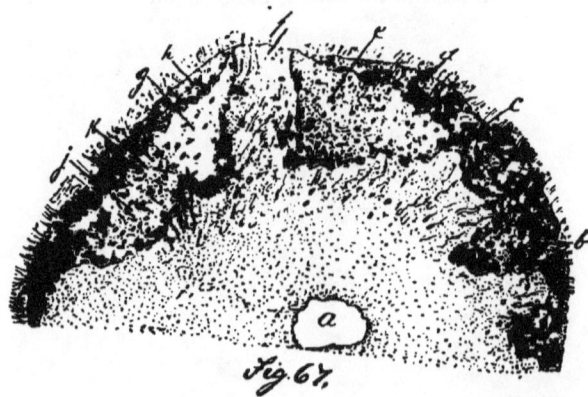

Fig. 67.

Fig. 65, ½ in. obj, One half of the apex of the root of a lower molar. From a dry section. a, Pulp canal. b, Dentine. c, Cementum. A number of absorptions have occurred at d. Absorptions have proceeded from the second lamella of the cementum and have penetrated the dentine to a considerable depth. These have been refilled with a somewhat irregular deposit of cementum. Along the line e, a very considerble absorption has cut away the entire apex of the root, removing not only the cementum, but evidently a considerable portion of the dentine as well. From the appearance of the incremental lines, this seems to have occurred contemporaneously with those pointed out at d. The exposed dentine has been again covered with cementum, which is fairly regular, though its incremental lines are not clear. f, An absorption that seems to have been in progress at the time of extraction. ,

Fig. 66, ½ in. obj. From a section of a bicuspid with its alveolus, showing a pit-like absorption upon the side of the root in which the redeposit of the cementum has begun. a, Dentine. b, Cementum. c, Peridental membrane. d, Bone forming the wall of the alveolus. e, Absorbed area of cementum. It will be noticed that a new deposit of cementum has begun the filling of the area, and that the soft tissue in the area of absorption is of a cellular type. The bone also shows the effects of absorption in the cutting away of portions of the rings of the Haversian systems at f, while at g the presence of osteoclasts shows that absorption is in progress at that point.

Fig. 67, ½ in. obj. Cross section of the immediate apex of the root of a cuspid tooth, showing large areas of absorption. a, Root canal. b, e, g, and j show extensive absorption areas that have been refilled with cementum, while c, d, h, and k show smaller absorption areas that have occurred later. Some of these areas show the included fibers of the peridental membrane plainly, while others do not, probably for the reason that the section is not parallel with them. At f, the original or regular deposit of cementum reaches the present surface. The plane of the section is not such as to show the incremental lines, and therefore the relation of the absorptions to these cannot be seen.

THE PERIOSTEUM AND PERIDENTAL MEMBRANE.

CHAPTER I.

PRELIMINARY.

In the study of histology there have been great advances within the last few decades. This advance has been along special lines to which attention has been strongly drawn by the results of individual effort, or in which the needs of the suffering public have directed investigation. Cohnheim and Stricker, with numerous colaborers, have done much to unravel and make plain the formerly mysterious tissue changes which occur in inflammation. By the investigations of numerous workers the tissue forms and modes of growth of the various tumors have been made so clear, that even the young worker in pathological anatomy may readily recognize their various forms, and classify properly those that may come under his lens.

The more difficult special tissue forms of the eye and ear have been so plainly unfolded by workers in these fields that it is no longer a question as to the forms of the elements found, but the discussion is carried to the domain of the more intimate and special physiological function of individual groups of cells which are recognized by all. The very complex structures of the brain, spinal cord, and the various ganglia of the nervous system have been searched so closely that the discovery of form elements yet undescribed seems almost impossible. And as it is with these, so it is with a large majority of

1

the form elements of the human body. Yet there are many special fields opening up for further discovery and waiting for laborers.

In this work each advance in discovery has brought with it a corresponding advance in *technique*. New and better means of bringing difficult and hidden form elements into view have been so rapidly brought forward that one who has been out of the work for but a few years will, on entering the histological laboratory of to-day, find himself confronted by apparatus and re-agents, much of which will seem new and strange ; and will find that these have modified the views of tissue elements with which he had been familiar, and have brought them into bolder relief, developing finer elements of structure than had been possible by the older methods of procedure. That which had been known becomes better and more intimately known ; and this better technique calls for re-working in the formerly worked out mines of histological inquiry for the finding of the finer grains of information missed by former laborers.

Each field in histology seems to be worked over and pondered over anew, as new pathological factors have fastened the attention of specialist or general practitioner. This is the case whether the new factor be based upon some new fact discovered, or theory propounded ; for each thought that gives promise of developing truth must be tried and tallied with the form elements with which it is associated.

It is considerations such as these that have prompted me to undertake anew the study of the periosteum and peridental membrane. Fifteen years ago I went over the subjects pretty closely, but at that time there seemed to be no special call for more definite information in regard to them. Upon the peridental membrane there had not been much written, and there did not seem to be much interest in the subject among the dental specialists.

Since then, however, attention has been strongly called
to the structure of this membrane from several directions
almost simultaneously, and an intense interest awakened.
These are, first, the efforts that have been expended in
the study of its destructive diseases, which was stimulated primarily by the late Dr. Riggs of Hartford; secondly, by the greater and more general interest recently
felt in the correction of irregularities of the teeth, in
which changes in this membrane and the relations of the
parts which it unites are brought about; third, by the
greater interest that has been manifested by the masses
of the dental profession in the retention of pulpless teeth,
and roots which have lost their crowns, and which are
dependent upon the continued health of the peridental
membrane under modified conditions; fourth, by the re-
vival under varied forms of the ancient methods of
replanting and transplanting teeth, the success of which
is supposed to be dependent, in whole or in part, upon
the reconstruction of the peridental membrane, or its reattachment to the teeth; and, fifth, by the singular fact
that has of late been noted more accurately; that a large
proportion of the teeth thus replanted with seeming success, are finally lost by absorption of their roots; a matter
which seems to depend upon some mal-condition of the
tissues of the peridental membrane.

All of these considerations call earnestly for an intimate knowledge of the histology and physiology of this
membrane as the basis for the formation of correct views
of its pathology and recuperative powers when subjected
to disease or serious injury. In order that I might obtain
correct views for presentation, I have gone back to the
tissues themselves for information and have made a re-
study of the subject *de novo*, availing myself of the new
methods of procedure, and preparing such a number of
sections from various sources as would seem to give
every possible view of the subject. In this study it has

not seemed judicious to confine my labors to the peridental membrane, either in the study or in the presentation, but to unite with it the study of the periosteum for
the purpose of having a broader field of comparison.
This is at once suggested by the natural kinship of these
tissues, and has been rendered the more necessary by
previous views that have been entertained as to the distinctions between them, or their identity. Indeed, the
relations of these membranes are such that the periosteum must be studied in order to arrive at correct views
in regard to the structure and function of the peridental
membrane, which will clearly appear as we proceed.
They are alike in many of their features while presenting
points of sharp dissimilarity, and by studying them together each becomes better understood. The task is an
unusually difficult one for several reasons. First, it is
impossible to obtain suitable sections for the examination
of these tissues without first decalcifying the bones and
teeth ; for we must study them in their normal relations
to the parts with which they are associated. The acids
to which they must be subjected in the process of decalcification are not without effect upon the tissues. This
is injurious in a large degree, and robs them of that
freshness so necessary to the gaining of good views of
their constituents.

Again, the selective stainings that are so valuable in
histological determinations depend upon the finer chemical qualities of certain constituents of the tissues, their
cells, or fibers, which cause certain of these to absorb a
dye or color while others do not, thus distinguishing
them. The necessary subjection to acids in decalcification disturbs these finer chemical relations seriously, so
seriously as to render the use of some of the finer staining agents unavailing, and causing much annoyance and
imperfection in the use of others.

Again, in the study of most of the tissues a little

shrinkage in the process of hardening for the purpose of making sections is of little or no consequence, for all being soft they will presumably shrink in the same degree, and their relations will not be disturbed; but in the membranes we are to study we have soft tissues combined with bone, and teeth and their mutual relations must be maintained. If shrinkage occurs in the softer portions these relations are disturbed and the object defeated.

CHAPTER II.

Before proceeding with this study it will be well to review the more elementary histology of the class of tissues to which these membranes belong; to first learn of what tissue elements they are mostly composed, and the character of these elements individually; and afterward we shall be enabled to study them more intelligently in their combinations and peculiar forms in special localities.

These membranes belong to what is termed the connective tissue group, and in structure are very nearly related to many other parts; so much so, indeed, that in most of the works on histology a description of the group as a whole has been considered sufficient without separate descriptions of the special membranes, or with a simple mention of some of the more important structural peculiarities. This would seem to be sufficient to the ordinary student of histology who often has the structures under observation, but it seems that to those who depend mostly on reading for their information in regard to such subjects, which up to the present time includes the greater number of both medical and dental practitioners, this does not serve the purpose when attention has been strongly called to a particular one of these. If one attempts to look up the special subject of the periosteum or peridental membrane in any or all of the current histological works, he will find the descriptions short and rather vague; indeed that the literature of the subject is very incomplete. Yet if these descriptions, short as they are, be taken together with a good practical knowledge of the histological characters of the elements of the group

6

of structures to which they belong, a comprehensive idea
of them will be gained. Still it must be admitted that
more especial description is needed in the light of the
recent interest awakened in the peridental membrane.
Furthermore, additional studies of the periosteum, especially from the pathological standpoint, are very much needed,
and these should be preceded by further studies of its
regional histological characters, especially differences in
the internal layer and the varying modes of its attachment to the bone. Previous studies of the periosteum
have related almost solely to its bone-forming powers
giving little or no consideration to the special elements
which serve purely physical functions or distinctions between these.

This large group of *fibrous membranes* is usually made
to include structures which, though seemingly widely
separated, are closely connected in their structural peculiarities. That is to say, though they may seem to serve
widely different purposes, or, better, are connected with
widely different organs, they are all emphatically *fibrous*
in their structure and differ only in the peculiarities of
their fibrous arrangement, in preponderance of the white
or elastic varieties, and in the number and character of
the cells which may be contained within the fibrous network. Again the purposes subserved by these different
fibrous membranes when closely studied, are found to be
as similar as their structure. They are all *coverings* for
other structures, and form their connections with neighboring parts, and, while in themselves they are indifferent
tissue, they are generally made subservient to functioning
tissues by conveying bloodvessels and nerves, and holding
in some part of their netword embryonal cells for the
supply of the needs of the tissues which they envelop or
connect.

DEVELOPMENT.

The connective tissues are developed in a soft transparent homogeneous material which has been known as

ground substance, basis substance, gelatinous substance or matrix. This material is in large proportion in the primitive or developmental state of this tissue, both in the fetus and in the early development of it as it occurs in the healing of wounds in the adult. Within this matrix the cells lie imbedded, and in this state it is usually termed gelatinous tissue. In this matrix the cells may exist in such great numbers as to obscure the ground substance, or they may be but sparsely distributed, and their development may be studied step by step as the adult tissues are assuming their forms, and through those changes by which certain of the fibrous elements are derived from them. In a few of the organs of the adult this tissue seems but partially developed, notably in the dental pulp (see fig. 16), the ground substance remaining in large proportion, and the cells being developed with but slight inter-mixtures of the fibrous elements. These have been termed myxomatous tissues. But generally the tissues undergo such development as to completely change their apparent character. The ground substance disappears more or less completely, giving place to fibers of various forms. Among these, two varieties, differing essentially the one from the other appear, known as the white and the yellow, or the inelastic and the elastic fibers. The former is in much the larger proportion, and its development is traced very directly from the primary cells found in the basis substance.

In fig. 1, I present an illustration of the tissue taken from beneath the epithelium of the abdominal wall of a human fetus in the sixth week, and in fig. 2, a specimen from the same locality from a little older fetus. (The umbilical cord is usually recommended for obtaining views of embryonic tissues.) In fig. 1, the cells are round, oblong or irregular in form and the cell contents slightly granular, and presenting either no clearly defined nucleus, or it appears but faintly, or at least it has not the prominence seen in the epithelia. In fig. 2, the cells are generally assuming a lengthened form in a common direction, and some of them present pointed extremities, yet no true fibers are present.

In fig. 3, the cells are illustrated in a more fully developed state, in which the points are drawn into long slender filaments. These may often be found in the subcutaneous tissue of the fetus, lying side by side and end to end with their filaments joined together, apparently, in such relative position that the full length of the filaments of two cells lie side by side, as in the two lower cells in fig. 3. Sometimes these fibers seem to be fused into one· As the development proceeds this appearance is changed by the development of numerous fibers between the cells; the cells meanwhile becoming smaller. As to the precise mode of the formation of these fibers there is still some difference of opinion among histologists. One view regards them as developed from the ground substance in the immediate neighborhood of the cell, and another, that the fiber is shed out from the cell itself—is the direct product of the cell. One can hardly trace this development as it proceeds without feeling a conviction that the fibers arise, at least, under the immediate supervision of the cell. I shall, therefore, call such cells fibroblasts. This fibrillation proceeds until the ground substance has disappeared and given place to a fibrous tissue which presents the appearance represented in fig. 4. As the tissue grows older the cells are separated more and more widely, and become smaller, until finally in the older tissues they are represented only by thin scales lying among the fibers. Many, and often almost all of them, disappear entirely. This disappearance is often quite complete in the more compact fibrous tissues, especially the tendons. In the developed tissue the fibers are generally not straight unless put upon the stretch, but pursue a wave-like course, as shown in fig. 5.

2

These fibers are very small, and are usually gathered together in bundles in which the individual fibers may seem to have but slight connection with each other, and form broad, flattened, wavy belts of loose texture, running parallel or crossing each other in various directions, as in the peridental membrane, or may be formed in close, compact bundles, that assume the form of large fibers (coarse fibers—fig. 6), running nearly parallel, but interlocking with each other, as in the outer layers of the periosteum, or may cross each other in every conceivable direction, leaving larger inter-spaces or meshes (areolæ) between them, as in the areolar tissue, or may be compacted into a dense, tangled mass, as in the gums. In some of these forms the fibers are cemented together into bundles by an intervening substance.

The individual fibers are never seen to branch or divide, and we may often search the tangled nets of the coarse fibers or bundles in vain to find them dividing, but in some localities these are found abundantly, as in the gingivæ. These branchings are, I believe, always effected by the splitting off of a portion of the fibers, which form a bundle, or coarse fiber, as represented in fig. 6 (from a silver nitrate preparation). The smaller bundles which split off in this way sometimes join and form a part of another coarse fiber, and by these divisions and junctions form nets through which fibers running in different directions may pass, or they may inclose cellular elements. This kind of branching occurs perhaps only in the softer forms of the coarse fibers. Occasionally, when the tissue is passing from a more firm to a looser texture, the more solid, coarse fibers are seen to break up and spread out in finer fibers, as represented in fig. 7.

These coarse fibers are seldom round. They are often seen to form bundles interlocking with each other ; often in compact masses. These, when seen in cross sections, present exceedingly irregular forms and sizes (fig. 8).

Fine sections of the tissue of the gums, when delicately stained with silver nitrate or osmic acid, give a great variety of views of these. In this position the coarse fibers are very closely interwoven, and every field will present cross sections differing in outline and configuration of the cut ends of the fibers.

In some positions, however, we find branching fibers of a different type. The cells, in their development, instead of assuming the tapered spindle forms, with processes at either end, present irregular star forms, sending out three or more filaments, as represented in fig. 11. In some positions these cells are seen to remain in this form without further fibrillation, as in the dental pulp; but in others, notably in the framework of the lymphatic glands, they form by their fibrillation an intimately branched network, as represented in fig. 9. These are known as *reticular* fibers, and form reticular tissue. In these the star-shaped cells are seen at the junction of the branches, and in the mature forms seem to lie upon them as a flattened scale, which may be removed by brushing.

These fibers, like those previously studied, are not round. The shapes, as shown in cross sections, present indefinite variation, with a tendency to elongated forms, as illustrated in fig. 10, showing the fibers to be irregularly flattened.

The development of the yellow or elastic fibers has not been traced so successfully as the white, and there is still much uncertainty regarding the manner of their origin. Krause regards them as being developed from cells in a manner similar to that of the reticular fibers, except that elastin is formed instead of the glue-giving substance of the white fibers. Boll and others have also traced their formation from cells. But a large number of those who have examined the subject have failed to trace the development of these fibers from cells.

Others have thought that elastic fibers are developed

by the formation of granules of elastin in the basis substance, and the union of these, end to end. The same
material is also found in the form of a very thin, elastic,
and apparently perfectly homogeneous membranes, which
are supposed, according to this view, to arise by the union
of the granules of elastin in the form of sheets, which become united into a continuous membrane.

The elastic fibers are found in almost all parts of the
soft tissues of the body, except the epithelial structures.
In another form this substance may be demonstrated here
also, serving as a connecting substance for the epithelium.
If a very thin section of the stratified epithelium be
treated with a 33 per cent. solution of caustic potash, by
the plan discussed later for the demonstration of elastic
fibers, the epithelial cells will gradually disappear, leaving a delicate net of elastic material, which accurately
represents the former junctions of the cells. A few ligaments are composed of elastic fibers almost as pure as the
ligamentum nuchæ of the ox, and ligamentum subflava of
man; but in most places they are but scantily distributed.
Wherever found, they present the same characteristic
form of network. The fibers divide dichotomously and
form junctions freely, and in this form are interwoven
with the white fibers of the areolar tissues and the fibrous
membranes; and wherever we find these of loose texture,
we find large intermixtures of the yellow fibers. I have
represented in fig. 12, a portion of these nets as seen at
the points of reflection of the mucous membrane of the
lips from the gums. It is a little unusual, in the fact that
it represents nodal points from which several fibers radiate, while usually they divide only dichotomously, as represented in fig. 13.

The point of reflection of the mucous membrane from
the gum tissue at the labial side of the teeth, is a very
good place to study them. Here the tissue is of a very
loose structure, and the mucous membrane is united to

the parts beneath by a fine network of elastic fibers. Carmine or hematoxylin stainings, mounted in glycerine, serve the purpose best when the networks are sought; but for examining cross-sections of these fibers as they occur entwined among the coarse white fibers, osmic acid or silver nitrate stainings, mounted in balsam, give better results (fig. 15). · In fig. 13, I have represented the fibers as they appear when teased out from elastic tendon.

Elastic fibers, like the other varieties, are seldom round. In fig. 15, I have illustrated these forms as seen in cross-section in a silver nitrate staining, in a group noticed passing between some coarse white fibers. Elastic fibers show a peculiar disposition to curl at the ends when cut or broken, which I have represented in fig. 14, giving examples taken from the position mentioned above. Here we will often see the short pieces cut off in thin sections, very much curled.

The cellular elements of the fibrous membranes, other than those already described, are mostly peculiar to the position, rather than to the fibrous tissue, and belong to the tissue invested rather than to the fibrous investment, and their consideration belongs to the regional descriptions. The leucocyte is found among the meshes of these membranes very generally, and other cells have been described as especially belonging to them ; particularly a round, nucleated cell larger than the leucocyte, and certain forms of the branched corpuscles. In some positions these latter appear abundantly, notably in the membranes of the eye. It seems probable that some of these are developmental types destined for some use in the neighborhood, rather than belonging specifically to the connective tissues as such. There are, however, young cells present undergoing development in perhaps all of the tissues of young animals, and possibly in the old, that, in the regeneration or augmentation of the tissue, pass through the phases already described. And it is also probable that many of the peculiar forms occasionally seen are cells that have become stationary, have begun to retrograde, or have, from peculiarities of environment, assumed modified forms.

Fatty tissue consists of connective tissue cells, filled with oil, which usually lie heaped together in little groups, or may form great masses by aggregations of these. It is most abundant in areolar tissues of loose texture, but is occasionally found in the fibrous membranes.

The fibrous membranes act very largely as a depot of supplies to the tissues which they invest. They bear the bloodvessels and nerves, and in some cases they receive, partially, as in the periosteum for the bones, and in others wholly, as in the peridental membrane for the cementum, the pabulum from the blood to be transmitted through their meshes to the point of assimilation.

The local characteristics of the individual membranes are continually modified by the deflection of their fibers this way or that, to give place for the passage of bloodvessels and nerves, and for the investment of them. These deflections are often of such a character as to mislead the observer if only one or two sections are examined.

In these forms, and inclosing in the meshes formed by its fibers, varying numbers and forms of cellular elements, this tissue is distributed throughout the body. It is continuous everywhere, and has been described by different writers under a great variety of names, according to the local peculiarity of the tissue and the positions in which it is found. It is found under the mucous membranes — submucous tissue; under the serous membranes — subserous tissue; under the skin — subcutaneous tissue; and about the bloodvessels it forms a continuous membranous sheath, or investment, and in this way gives them support and protection. In the same way it forms the investment of the nerves — neurilemma; and incloses each muscle in a distinct sheath — myolemma; and dipping in between the muscular fibers, surrounds each one individually — sarcolemma; and serves to connect them with their tendons, or with the periosteum. It invests the glands, holding their lobes in position, and, following the ducts into the substance of the gland, forms an investment for each lobule, and within this substance the bloodvessels that supply the gland ramify. It forms the support for the organs of the hollow viscera — peritoneum — pleura; it invests the brain — dura mater — arachnoid membrane, and forms the investment or matrix for its functioning cells — neuroglia; it incloses the heart in a closed sac — pericardium — and forms the investment of the eye — sclerotica. In strong membranous sheets — fascia — it binds down the muscles, and holds them in position; it forms the investment of the bones — periosteum — and serves to attach the roots of the teeth to their alveoli — peridental membrane. In a still more condensed form in which the fibers lie parallel with each other, it forms the tendons which connect the muscles with the bones, and the ligaments which connect the bones together.

This tissue also stands in very close developmental relations with still other and seemingly very different tissues. Cartilage and the bones are developed directly from a connective tissue matrix, and seemingly the one is developed from the other, though close examination seems to reveal the fact that the development of the one displaces the other wholly or in part. The bones, at least, are developed from specialized cells — the osteoblasts, which seem endowed with a special bone-forming power. These cells are, however, developed from connective tissue cells, or at least from the cells that, from all that has as yet been learned of them, are the same as the embryonal connective tissue cells. This point will be exam ined more particularly later.

It appears also from the teachings of comparative histology that one of these tissues may be substituted by another of equivalent value. That which in one animal appears as ordinary connective tissue may in another be of quite a different reticular type ; and that which is represented by cartilage in one may be substituted by bone in another. These changes are often remarked also in the developmental stages of the same animal. Thus in mammals the greater part of the skeleton is first represented in cartilage, which is afterwards replaced by bone, while a few of the bones, as the cranial, are first represented as fibrous membranes, in the midst of which the bones are formed. This is called the inter-membranous formation of bone. Three modes of the formation of bone are usually recognized ; the inter-cartilaginous, inter-membranous, and the sub-periosteal.

Further than this these tissues have other relationships from a physiological point of view. Their significance in the action of the healthy body is of a more subordinate kind ; though they make up an enormous proportion of it. They represent, as is usually said, tissues of lower vital dignity (Frey), and seem in a degree subordinate to the more proper functioning tissue for which they form an extended framework in the meshes or cavities of which the muscles, vessels, nerves, gland cells and organs lie imbedded.

The name then of connective tissue seems to be entirely appropriate. If we farther reckon the muscular tissue with this group, which seems proper according to most histologists from its developmental relations — though it is widely specialized, — it may aptly be termed the *tissue of support and motion*, while the tissues of the epithelial type constitute the *tissues of function and protection*.

CHAPTER III.

Perhaps it would be well, before going farther, to indicate · briefly the methods I have employed in the preparation of the tissues from which my studies and illustrations of those to be described have been made. These have been taken, for the most part, from young and small animals, such as the cat, lamb, pig, dog, etc. The human fetus, and also tissues taken from the adult, have been employed in sufficient amount to make reasonably good comparisons, but the difficulty of obtaining these sufficiently fresh to give the best results, is quite obvious. The time between the death of the animal and the immersion of the tissue in the fluids by which it is prepared for cutting, should be counted by minutes, never by hours. For this purpose I have used Müller's fluid and chromic acid, giving the preference to the former, and have usually added the acid for decalcification, after the first day. It has been considered important that very small bones be used, in order that the time of exposure to acids in the process of decalcification be as short as possible. It is exceedingly difficult to cut large bones into sufficiently small pieces, without disturbing the relations of the soft portions of the tissue, especially in the loosely attached portions of the periosteum. The injurious effect of acids has been closely studied, and it has been found that the element of *time* is, within certain limits, more important than the strength of the acid solution employed; that is to say, a tissue decalcified in one day with a three per cent. solution of nitric acid, and then thoroughly de-acidulated by copious ablution, will come under the lens in better condition than if exposed to one-half per cent. for five or six days.

The use of alcohol in hardening has been avoided as far as possible, on account of the shrinkage which it induces. Indeed, it has been limited to the dehydration of the tissue for the purpose of impregnation with the imbedding material, in cases in which this is demanded. Some sections of each class of tissue have been made without dehydration, or any form of impregnation, for purposes of comparison and the formation of conclusions as to the changes induced by the different materials used for these purposes. By this mode we are unable to obtain sections sufficiently thin and regular for general study, but may obtain scraps that will reveal the tissue characters for comparison. Such studies demonstrate that all of the modes of impregnation yet used for the purpose of cutting sections, injure the tissues in some degree, and that this injury is very closely associated with the *time* the tissue is allowed to remain in the imbedding material. Of these I have used gum arabic, celloidin, bayberry tallow (the concrete expressed oil of *Laurus nobilus*, or bayberry tree), paraffin, and paraffin modified by additions of cosmoline. The finest sections may be cut in paraffin. The disposition of this material to curl up before the knife, is readily avoided by laying a piece of fine tissue paper wet with alcohol upon it, and cutting under this. This paper also serves well to transfer the sections to fluids. The necessary subjection of the tissue to alcohol for dehydration, then to warm chloroform in the impregnation of the tissue, and the removal of the paraffin after cutting, is not without evil result. In this respect the bayberry tallow is better, as the use of chloroform is avoided, warm alcohol being sufficient both for impregnation, and solution and removal of the tallow after cutting. Each of these have given me better results, as to the final condition of the tissue, than the celloidin; but, in order to obtain good results, the *time* the tissue remains in the imbedding material must only be counted by minutes, never by hours; and if, after its removal, and placing the tissue in water, it does not swell out to its normal proportions in every part, which failure may be detected after sufficient experience by its appearance, it should be cast aside and a new effort made. The gum arabic method is suitable only for very small bits of tissue, on account of the *time* necessary for hardening large masses. Tissue may be bound up, or shrunken in any of these processes for a very short time, without losing its resiliency, or power of resuming its original condition, but if it is continued beyond a certain time, this is lost in a greater or less degree. Careful additions of acetic acid to the water will often assist in the restoration of the normal condition of the tissue. The ordinary microtome has been used for cutting sections. The staining has been by carmine in its different forms, Picrocarmine, hematoxylin, osmic acid, and chloride of gold. Double stains of carmine and hematoxylin, and pigmenting (see below) have been made use of. Sections from each portion of tissue cut have been studied, mounted in glycerine jelly, plain, also plain after acetic acid, and in each of the above stains, and then these studies repeated in similar mountings in balsam. The aniline dyes have been tried in tissues in which acids have been used to decalcify bones, but have not given me good results.

Pigmentation should be explained, as I do not know of the use of the process by others. It is done in two ways, making a diffusive or selective pigmentation as desired. Place the sections in osmic acid solution (one per cent.), and let them remain from half an hour to an hour. (1st.) Transfer to distilled water for half a minute, just long enough to remove the osmic acid from the surface, and at once place in a solution of hematoxylin, as prepared for staining—a thin solution is best—and allow them to remain until they have assumed a deep smoke or soot color, which will require but a few minutes. (2d.) Wash thoroughly in distilled water from half an hour to an hour, then transfer to solution of hematoxylin as before. The change to the soot color will be a little slower. Any purple color acquired from this solution may be removed by acetic acid without affecting the pigment. The sections may now be prepared and mounted in any manner desired, and will be found very transparent to transmitted light, provided the pigmenting has not been carried too far. In (1st) the pigmenting will effect all the tissues alike, is diffusive, but in such a way that all of the elements come fairly into view. . In (2d) the pigmenting is selective, the osmic acid resisting removal by water is reduced as pigment by the hematoxylin. This pigmenting rests on the fact that a mixture of osmic acid and hematoxylin throws down an amorphous black deposit, and this is obtained in the tissue in such a fine state of division as to resemble a stain when the highest powers of the microscope are used. Some portions of the tissue hold the osmic acid—at least do not give it up to water very readily—hence the selective pigmenting of the tissues that are well washed after removal from the acid, before being submitted to the hematoxylin. In this way, ordinary epithelium may be made to resemble natural pigment cells, the cell body being pigmented deeply, while the nuclei and cementing substance remain transparent.

A word as to the illustrations. These are all made from tissues freshly prepared for the study of this subject, and are done with as much care as to accuracy of representation as I have been able to bestow. The manner of the representation of the tissues generally employed, is in a large degree conventional, and my illustrations are no exception to the rule. That which I have made out to my own satisfaction, I have endeavored to represent clearly, avoiding the representation of either shadows or suppositions. I therefore make no claim that the pictures are *exact* representations of individual fields in my sections, but are rather what I make out to be the actual forms of the tissue elements and their relations to each other, after having made the best study of them that I am able to do at the present time.

CHAPTER IV.

THE PERIOSTEUM.

The periosteum forms the immediate covering of the bones. It is continuous at all points except those surfaces covered by the articular cartilages and the attachment of the ligaments and tendons.

It is not, therefore, continuous from bone to bone, except in those united by suture, as the cartilages mentioned uniformly clothe the ends of those united by joints. Each of the long bones, and most of the short ones also, has its individual periosteum, which encloses it as in a sack, and is closely adapted to all parts of its surface.

If the flesh is carefully removed from any of the long bones the periosteum will be seen to present a smooth, white, lustrous appearance, much like the surface of a tendon, over a large part of the surface, but at certain points which correspond with the attachment of muscles, or fascia, it will be left more or less ragged and dull, for at such points the superimposed tissues are firmly adherent and must be cut away with the knife. At all other places the tissues separate from it easily and smoothly, indeed, are not attached, or are attached only by a very slight network of reticular or elastic fibers which break away readily and, to the naked eye, leave no sign of their presence. If now we slit up the periosteum lengthwise the bone, along a smooth portion, and insert the handle of the scalpel beneath it, it will be found readily separable from the bone over the greater part of its surface. Indeed, the attachment seems to be but little more intimate than was that of the tissues to the outer surface. However, if the

22

detachment be closely followed it will be seen that at many, or perhaps only a few points, fibers adhere to the bone, and are broken. These are, in the main, very small blood vessels that enter the bone from the periosteum, but occasionally a few fibers of the periosteum enter the bone also.

In the progress of the detachment a point is arrived at finally where this easy separation ceases abruptly, and the periosteum becomes firmly adherent to the bone. It is now found, in the effort to continue the detachment, that the periosteum is a very thin, tough, inelastic membrane that is torn with difficulty, but it is impossible to continue the separation from the bone otherwise than with the knife, and the extreme thinness of the membrane renders this difficult. An examination of these adherent points reveals the fact that they are, first: points at which some of the tissues are attacked to the outer surface of the periosteum, as muscles or fascia; second, near the ends of the bones where the periosteum approaches the articular cartilages; third, wherever it approaches the insertion of tendons or ligaments; fourth, wherever mucous membranes, or the skin, seems adherent to the bones beneath, as at the entrance of the meatus auditorius, the gums, mucous membrane of the nose, etc. At all such points the periosteum is as firmly adherent to the bone as if it formed an integral portion of it, and serves as the medium of attachment for the superimposed tissues. Through this medium many attachments of muscles, fascia, etc., are effected, and these points of attachment will intercept and prevent the separation of the periosteum from the bones at many points. This feature of the anatomy of the periosteum has not yet been studied in detail. Yet its importance in the management of diseases of the bones, especially the suppurative diseases, when pus is likely to find its way beneath the loosely attached periosteum, must be apparent to every surgeon. While I can not now undertake this part of the subject *in extenso,* I propose on another page to consider very closely the character of the attachments of the periosteum at different points.

Histologically, the periosteum is composed of fibrous tissue, in the meshes of which are found certain cellular elements. It presents for examination:

1st. An outer layer of coarse white fibrous tissue.

2nd. An inner layer of fine white fibrous tissue.

3rd. Elastic fibers.

4th. Pentrating fibers, or fibers of the periosteum that, in the growth of the bone, are included in its substance. (Fibers of Sharpey.)

5th. Osteoblasts, or a layer of cells that lie between the periosteum and the bone.

6th. Osteoclasts—cells that absorb bone.

The white fibrous tissue is everywhere disposed in two layers, an inner and an outer; or a layer of coarse fibers forming the outer portion, and a layer of fine fibers forming the portion next to the bone. The yellow or elastic fibers are found mostly intermingled with the coarse fibrous layer. They are usually very difficult of observation, and do not, as a rule, appear in sections as ordinarily prepared.

OUTER LAYER.

The size and arrangement of the coarse fibers in the formation of the outer layer is exceedingly variable in different regions of the osseous system. On the long bones they are generally smaller than upon the short, while I have found the largest fibers about the bones of the face. The rule is that the periosteum, as a whole, is thicker and stronger at exposed points where the bones are near the surface, and is more delicate when deeply covered with other tissues. Hence we find it thin, and its fibers correspondingly delicate on the shafts of the long bones, especially such as the femur, humerus, etc.

In these positions the coarse fibers of the outer layer are small, and for the most part run parallel with the long axis of the bone. (See fig. 17.) The fibers are usually very much flattened, and the fine fibers of which they are formed, not very firmly bound together. Indeed, they are often disposed in ribbon-like layers, with the flat sides horizontal to the surface of the bone, and the edges of these are often joined in such a manner as to form a continuous sheet of fibrous material. This is especially the case when the periosteum is deeply covered with muscles which perform sliding motions on its surface. In such places this portion is often made up of a number of lamellæ thus formed, which are very loosely joined together, so that by careful manipulation it may be separated into a number of complete lamellæ. The fibers which constitute these layers do not all run in the direction of the long axis of the bone, but some are interposed which cross these at right angles, or in the direction of the circumference of the bone, as shown in fig. 18, in a section cut lengthwise, from the tibia of the pig. This example shows five layers of circumferential fibers, and those marked f and i, have shifted from their position in mounting the section in such a way as to present the sides of short sections to view, instead of the ends, and serve well to show how the fine fibers are joined into ribbon-like forms. The figure, as a whole, illustrates how readily the different layers are separable, though, as combined, they are calculated to give great strength, at the same time accommodating sliding movements readily.

I have endeavored to represent every portion of it just as it happened to lie in the preparation. This may be regarded as an example of the more complex arrangement of the coarse fibers, or outer layer, in the non-attached periosteum. The disposition of the fibers is usually much more simple, presenting fewer running in a circumferential direction until, finally, none whatever can be found.

4

This simpler form I have represented in fig. 17, from a lengthwise section from the femur of a kitten. Every gradation between these may be found. For this illustration a point has been selected where the outermost fibers have been broken by the needle in detaching the superimposed tissue. Some of the fibers beneath are also a little separated, and in the central part the layer of osteoblasts is pulled partly away from the bone, displaying their processes to advantage. It will be seen that the fine fibers, a, cease abruptly, giving place to the coarse fibers of the outer layer c. By comparing this illustration with fig. 21, and noting the difference in the size of the osteoblasts (for the illustrations are drawn with different powers) some idea will be gained of the difference in the size of the coarse fibers in different regions. In drawing fig. 17, the $\frac{1}{12}$-inch immersion lens was used, while in fig. 21 the $\frac{1}{4}$-inch dry was substituted. These fibers (fig. 17) are round or irregularly flattened, and show none of the ribbon-like forms seen in fig. 18. It is the form of this layer most commonly met with on the shafts of the long bones, though the gradations between these two figures are sufficiently common. In both of these figures I have illustrated the delicate reticular tissue by which the periosteum is very loosely attached to the superimposed parts.

As the ends of the bones are approached the periosteum is thinner, and often the coarse fibrous layer is found lying almost flat on the bone, most of the inner layer having disappeared, and at many points the osteoblasts are not to be seen. (Fig. 19.) At frequent intervals, however, sometimes continuously for a space, osteoclasts (*f. f. f.*) have taken their place, and are trimming down the surface of the enlarged ends. In this region the fibers of the periosteum enter the bone and in this way form the firm attachments noticed at their ends. (Fig. 19.) This happens to such an extent that, in pursuing the study of sections cut lengthwise the shaft of the bone, up to the articular cartilage, one is impressed with the idea that the whole of the periosteum has sunk beneath the surface of the bone. As this occurs the fibers of the outermost parts of the coarse fibrous layer often seem to unite into a fibrous sheet which is inserted finally at the margin or fringe of the articular cartilage, or into the cartilage itself.

On the bones of the face and other positions where the periosteum lies near the surface of the tissues, the outer layer is composed of very large, white fibers, with which a small quantity of yellow elastic fibers is mingled. The white fibers form an intimate network, being closely interwoven with each other. (See figs. 20 D, and 21 E.) Sections cut in almost any direction will show longitudinal fibers, but a disposition to run in the direction of the pull or strain of muscles or other tissues attached to the periosteum may be seen ; otherwise the direction will have a tendency to follow the long axis of the formation of the bone. In either case a considerable number take a transverse or a diagonal direction, passing through the meshes formed by the principal fibers. None of these fibers take a direction perpendicular to the surface of the bone, but they are so disposed that a fiber that may be on the inner surface of the layer at one point, may at a little distance arrive at the outer surface. In this way the fibers seem to be plaited together. (fig. 21 E), sometimes in a very compact layer only a few fibers in thickness, and sometimes the fibers are so disposed as to form several lamellæ, held together by occasional, or it may be very frequent, passage of fibers from the one layer to the other, or by a network of elastic fibers only. (Fig. 22.) Usually, even in cases of considerable thickness of this layer, coarse fibers may be traced that in their longitudinal course gradually approach one or the other surface.

The thickness of this layer is very variable. Occasionally it is only the thickness of two or three coarse fibers superimposed on each other; rarely the coarse fibrous layer is condensed into a single membranous sheet, to which the overlying tissues are attached. On the other hand, I have seen the thickness of $\frac{1}{4}$-inch in the human subject; however, I am not sure that the latter was entirely normal. The thinner portions are often those to which muscles are attached. Indeed, the statement is made by Krause (Allgemeine und Microscopische Anatomie, p. 68) that this layer is sometimes wanting at the points of attachment of muscles. Although I have made many cuttings through such points, I have never found this layer absent except when the muscle was attached to the bone by well-defined tendon, in which case none of the elements of the periosteum whatever remain, but the fibers of the tendon pursue their course uninterruptedly into the surface of the bone.

The characters described above are present in the coarse fibrous layer of the periosteum wherever it is found, without exception. At a few points it is blended with other fibrous tissues, especially with the mucous membranes and skin, as it is seen in the gums, and at the entrance of the opening of the external ear, and other points at which the skin is rigidly adherent. In these cases, if we proceed from the surface of the bone outwards, the first coarse fibers are always disposed as in the periosteum at other points, i. e., lying horizontal to the surface of the bone ; but after passing a few of these, the horizontal direction of the fibers is sometimes gradually, sometimes abruptly, lost, and the character of the tissue changes to the tangled fibrous forms of the skin, the gums, areolar tissue, or whatever may be the superimposed fibrous tissue.

Often, however, the periosteum remains entirely distinct from the superimposed tissue, and is united with it only by a scanty network of elastic fibers which allow of free sliding motions of the one tissue upon the other. All grades of connection, from this latter to the intimate commingling of the coarse fibers, may be found in different places. Many of the smaller muscles, and larger ones that have their attachments by a broad base, are attached directly to this layer of the periosteum. In case of the muscles, the sarcolemma of each individual muscular fiber is attached directly to these coarse fibers. (Figs. 20 and 21.) In a few cases fine fibers may be seen traversing the layer of horizontal fibers of the periosteum in a perpendicular direction, seeming to be condensed extensions of the sarcolemma as shown in Fig. 21. Occaionally these pass entirely through the coarse fibrous layer, and then appear to be continuous with the fibers of the internal layer. The fasciæ are attached to the periosteum by their fibers blending, or becoming continuous with those of its external layer.

INTERNAL LAYER.

The internal layer is of an entirely different character from the outer, both in the nature of its fibers and in their arrangement. It also presents great diversity of arrangement. In the consideration of this layer it will be convenient to divide it into *attached* and *non-attached*, as it presents notably different characters in its fibrous structure, and in the relation of its fibers to the bone which it clothes.

The *non-attached inner layer* of the periosteum is separated from the bone almost completely by an intervening layer of polygonal or flattened cells, the osteoblasts. (Figs. 17 and 18). None of its fibers pass into the bone ; while in the *attached* periosteum those of the inner layer do pass into it, or seem to spring out of it. (Figs. 21, 23 and 24.) It is composed of the finest and most delicate white connective tissue fibers, with which there are no coarse white, or yellow elastic fibers asso-

ciated. On the short bones these fibers seem not to be
disposed in any particular direction, or upon any specific
plan that I have been able to detect. They decussate
freely in every direction. On the long bones the fibers
of this layer are more generally parallel to the long axis
of the bone, as illustrated in fig. 21. though not universally so. In all young animals these have in their
meshes a considerable number of young connective tissue
cells in various stages of development, in addition to the
fusiform nuclei of the white fibrous tissue or fibroblasts.
In the main the fibers lying next to the layer of osteoblasts have a course horizontal to the bone; but in the
short bones, or in the neighborhood of attachments, this
is changed to a direction more inclined towards the coarse
fibrous layer, and the particular band or group we attempt
to follow will become intermingled with others and lost.
At another point immediately adjacent, the fibers are
seen cut across either directly or diagonally; but even
among these will be seen those that are horizontal to the
plane of the section. While there seems to be no uni-
formity in the direction of the fibers, the fibrous appearance is maintained, giving the impression of an intimate
intermingling rather than of a network. This appearance
may be much modified by the manner in which the section is prepared for observation. If it be with a good
selective stain this layer in young animals may appear
distinctly cellular, the fibers being much hidden, while
the cells are made prominent. If on the other hand the
fibers be rendered prominent by diffusive carmine staining, osmic acid, or pigmenting, the tissue will give the
impression that it is almost wholly fibrous. The various
plans of preparation should be employed in its study.
The fibers do not seem to branch and anastomose as in a
net, but rather to decussate with the utmost freedom,
rarely forming groups or bands of any considerable number running in a common direction. However, it is

apparent that in the portion next to the layer of osteoblasts they are more inclined to a direction horizontal to
the surface of the bone; while in the portions next to
the coarse fibrous layer their general direction has
become perpendicular, or more or less inclined to the surface of the bone.

In the long bones the fibers of this layer in the non-
attached regions very generally lie horizontal to the surface of the bone throughout its thickness, as is shown in
figs. 17 and 18, and run in the direction of its long axis.
The tissue is loose in texture and somewhat embryonal
in its character immediately adjacent to the layer of osteoblasts, but becomes more prominently fibrous as the bone
is receded from. Its attachments on either side are very
loose and easily broken up — so much so that it is difficult to keep the parts in position while mounting the
sections. In those sections in which the relations of the
parts are a little disturbed by spreading apart, we often
obtain the best displays of the tissue elements, and it can
be seen that the osteoblasts have processes which pass in
among the fibers lying next to them, and also into the
bone, forming a sort of attachment to it, which is, however, very easily broken up. This is well shown in figs.
17 and 18. This latter form is common to the shafts of
the long bones, and is almost universally present in the
non-attached regions, which may in general be expected,
except in the region of the attachments of muscles, fasciæ,
ligaments, or the approach to the ends of the bones.
In such positions the form of the attached periosteum is
assumed.

In the *attached portions of the periosteum* the fibers of
the internal layer exhibit a definite arrangement. This
presents certain variations at different points, but these
are only modifications of a definite plan. Here the fibers
are not separated from the bone by the layer of the osteoblasts, but *spring directly out of the bone itself,* and the

osteoblasts are seen to be disposed between the fibers, as
in figs. 20, 21, 23 and 24. Perhaps a more correct statement would be that the fibers spring out of the bone
between the osteoblasts.

At some points the former statement would be the
more correct, for the reason that the fibers occupy the
greater amount of territory, so that the osteoblasts are
crowded into various forms to accommodate them. Every
grade, from an occasional fiber passing out of the bone
between the osteoblasts, to an increase in numbers and
size which represents the insertion of the tendon, in which
no osteoblasts are present between the fibers may be found.

In the attached portions then, the fibers of the inner
layer of the periosteum spring directly out of the bone.
In order that this may be well seen it is absolutely necessary that extremely thin sections be cut parallel with the
fibers as they emerge from the bone, and in general this
will also give a good view of the arrangement of the
fibers of this layer of the periosteum, for the fibers pursue the same general course until they reach the inner
surface of the coarse fibrous layer. Each of these fibers,
after passing out of the bone, or immediately after rising
above the osteoblasts between them, breaks up into a
tuft of very fine fibers; indeed, in many sections it is
shown that that which in the main appears as rather a
coarse fiber as it makes its exit from the bone, is really a
compact bundle of very fine ones. These, on separating,
spread out fan-like, and intermingling and decussating
freely with others, take their way perpendicularly, or
inclined somewhat to the surface of the bone, to the inner
surface of the coarse fibrous layer to which they are
attached. The arrangement of the fibers as they pass
from the bone to the coarse fibrous layer varies greatly in
different positions. The most common form seen is that
in which all of them pass at more or less inclination to
the perpendicular, and join the coarse fibrous layer at an

acute angle, as shown in fig. 21. Yet every angle from about 45 to 90 degrees may be met with. Occasionally, however, we see them joining the coarse layer in inverse directions, decussating with others as shown in fig. 20. It is quite rare that the fibers join the coarse layer at right angles. In many instances their decussation in this layer is much more limited, and they pass quite directly from the bone to the coarse layer, forming a very regular sheet of fibers that are almost parallel. This layer has no elastic fibers, or at least they must be rare. Some observers state that these are found here, but I have repeatedly made the examination in the manner detailed below, without finding them.

ELASTIC FIBERS.

Elastic fibers form a network in the coarse fibrous layer that is very difficult to see without special preparation. This is partly on account of the fineness of the fibers themselves, but more especially owing to their relations to the coarse ones. In fine sections stained diffusely with carmine they may be imperfectly seen as white lines, but they are studied to best advantage by dissolving out the white fibers on the stage of the microscope. If this is done with sufficient care their arrangement can be quite accurately made out. This is done as follows: Place the section on the slide in water, lay on a cover glass, and carefully dry the slide at its edges; now fasten the cover securely at two points, preferably next the edges of the slide, with a little gutta-percha dissolved in chloroform, with balsam, or with wax. Now having placed the slide on the stage of the microscope in such a position that it will be inclined from end to end (it may lie flat if preferred), lay on a piece of blotting paper cut to fit the circle of the cover glass (a square cover glass may be used), and lay it on the highest end of the slide in such a position that it will touch all of the higher edge of the

5

cover glass not covered by the gutta percha. Also lay a piece of blotting paper on the opposite end of the slide, so that it will touch the margin of the cover glass. Here a central point of contact is sufficient. Thus prepared, saturate the upper bit of blotting paper with a strong solution of caustic potash (33 per cent. is best). This will gradually pass through under the cover and be absorbed by the paper below. A fresh drop should be added every few minutes continuously for several hours. The white fibers will first swell and become more transparent, and the elastic fibers meantime will come into view. Finally all of the white fibrous tissue will slowly melt down and disappear, and the only tissue left on the slide will be the elastic fibers and some remains of the bone.

This process may be checked at any stage by substituting distilled water for the potash solution, and if this is followed by glycerine, and glycerine jelly, a permanent mount of the object can be effected. It must be borne in mind that the solution of the tissue can not be stopped at once, and the particular stage desired for the preparation must be anticipated. The washing with distilled water must be continued for a considerable time to remove all of the potash.

If the process of the solution of the tissue be closely watched it will readily be discovered that the elastic fibers form a network in which the coarse fibers of the periosteum are inclosed, or that the elastic fibers are entwined about the white in such a manner as to prevent their separation, or, if they are somewhat separated by a strain, will bring them back by their elasticity. I have illustrated such a network in fig. 22, taken from a section from the lower jaw of the same series as that represented in fig. 21. They are not uniformly distributed in the coarse fibrous layer, but seem to be most plentiful where muscles are attached to a rather thick outer layer, and the regions of the attachment of the mucous membranes.

Along the shafts of the long bones I have usually found very few, and these seem not to penetrate the periosteum deeply but are, indeed, unusually joined to its surface, and serve to make a very loose attachment of the superimposed tissue.

The inner layer of fine fibrous tissue of the periosteum is generally destitute of elastic fibers. Only once have I seen a few of these penetrating to the surface of the bone. Frequently I have seen a few fibers passing some distance into this layer, but generally they are confined to its outer margin.

The blood-vessels of the periosteum are quite numerous, and present considerable variations in different regions. On the shafts of the long bones the larger vessels usually run in a direction parallel to the long axis of the bone, and lie between the periosteum and the superimposed tissues, or on the surface of the periosteum. These branch laterally, and anastomose in such a manner as to form a tolerably continuous network. This network receives here and there branches from the superimposed tissues. In some situations, especially in the attached portions, this network lies immediately beneath the coarse fibrous layer, or in the outer part of the internal layer, in many instances as nearly between them as is possible. (Fig. 21 D.) However, in those situations in which the coarse fibrous layer is thickened by the formation of two or more lamellæ, the network of blood-vessels is often found between these, a circumstance which has given rise to the statement by various authorities that the blood-vessels of the periosteum are found mostly in the outer layer. In my observation there has been much more irregularity in the blood-vessels of the periosteum of the short bones, which, I may say, would naturally be expected, both as to the position of the individual layers, and the regularity of the network formed. From the network of vessels thus formed in any of these positions frequent capillary branches are given off, also occasional larger vessels, which pass down through the fibers of the internal coat and enter the Haversian canals of the bone. In the attached forms of this coat, these branches very generally follow the direction of the main trend of the fibers of this portion of the periosteum, and in a few localities they are quite numerous, especially about the bones of the face, and notably over the surfaces of the alveolar processes. In the portions of the periosteum, with which the fibers of the mucous membranes, or the skin, are intimately blended, the position of the blood-vessels is notably irregular; indeed, they seem to pertain rather to the superimposed tissue than to the periosteum, and send frequent branches through the latter to the Haversian canals of the bone.

Occasionally I have noted a plexus of vessels in the internal layer very close to the layer of osteoblasts, but these are very small and infrequent.

The nerves of the periosteum are generally few in number; however, a considerable number of the larger vessels are accompanied by a small bundle of nerves, which are probably distributed mostly to the blood-vessels themselves. They enter the bones with most of the larger branches of the blood-vessels. At some points, nerves passing through the periosteum to enter the canals of the bone are very frequent. These are points where the nerves are required by organs situated within the bone. The supply of the peridental membranes renders them frequent in the periosteum of the alveolar processes.

CHAPTER V.

The cellular elements of the periosteum consist of developing connective tissue cells destined to form osteoblasts, osteoclasts and fibroblasts. The fibroblasts are such as are destined to reconstruct or augment in numbers the fibers of this membrane and have been sufficiently considered. However, I may say that a considerable number of connective tissue cells are found that seem not to show specific character. They seem not to be proceeding regularly to the development of fibrous material nor to be allying themselves to either of the other two forms, and are probably cells that have missed their destiny and therefore have developed irregularly. By careful search a variety of such may be found. They are mostly round or oval nucleated forms, but occasionally irregular star-shaped forms present themselves. Such cells seem to have no function to perform in connection with this membrane or in the locality in which they are found; they are not sufficiently numerous and regular in their distribution for me to suppose that they perform some undiscovered function which renders their presence necessary. Hence the supposition that they are cells which have missed their destiny and finally disintegrate and disappear. In the progress of our study of the other cell-forms, the function of which is obvious, we shall find sufficient evidence of faulty action, or of over-activity in certain directions, which is yet within the range of what may be termed physiological errors on the part of the elementary forms. These are errors in direction of growth, or removal of tissues, which are checked before

they become so pronounced as to be regarded as pathological; as, for instance, in the case of absorption of bone
beyond the needs of the time and its reconstruction afterward. By English writers the osteoblasts are usually
reckoned as belonging to the periosteum, while some of
the German authors have classed them as belonging to
the bones, and designated them as the cambium layer.
So far as they are connected with the periosteum at all,
their place is between the periosteum and bone in the
non-attached forms, and upon the bone between the penetrating fibers in the attached forms. However, the number of embryonal cells that are found among the fibers of
the periosteum in the immediate vicinity of the bone,
gives the impression that this portion of the tissue is the
place of the development of the osteoblasts, and that these
cells are destined to become such. The most plausible
supposition is that these embryonal cells are leucocytes
that have wandered in here from the blood streams, not
by any manner of chance, as this expression might indicate, but through the control of some unseen power which
causes these cells to congregate where they are needed
for building up of new tissues, or the repair of injuries to
the old ; and to develop into the necessary forms for this
purpose, whether it be for the formation of fibrous tissues,
the formation of bone, the absorption of bone, or for
whatever else may be needed which is in the power of the
connective tissue cell to perform.

The osteoblasts are polygonal cells which lie upon the
surface of the bone and usually clothe it as epithelium
clothes the mucous membranes. They vary greatly in
size, so much so, indeed, that no measurement will give
a very accurate idea of them. They are also placed very
differently in relation to the bone in different positions
and under varying conditions. In case of young bones
that are rapidly growing they are often very much crowded
together, and thus compressed into a great variety of

forms. Occasionally they are very much elongated, as, for instance some of the cells in fig. 25, taken from a cross section of the tibia of a young kitten. Here it will be seen that some of the cells reach the bone only by extending a process-like elongation between the neighboring cells (a) while others seem to be flattening down upon the surface (b). In my studies it has seemed to me that only those cells which are attached to the bone should be considered as osteoblasts. They are undoubtedly developed from the embryonal cells of the neighborhood, but it is not until the time of their attachment to the bone that their destiny can be definitely determined. Therefore, we can hardly say that more than a single layer of these cells is ever found upon the bone in any case. More than one layer is often made to appear by cutting sections diagonal to the surface of the bone.

The more usual forms of the osteoblasts appear in figs. 17 and 18, where there are not too many to conveniently cover the surface; and by the slight shrinkage that is almost inevitable in histological preparations, they are made to stand slightly apart. The processes of these cells appear prominently in figs. 17 and 18. These are very difficult of observation, and it is only under especially favorable circumstances that they appear; however, they are seen so frequently in favorable positions as to lead to the supposition that all osteoblasts that are so far developed as to come to lie upon the bone possess them.

These cells are found also lining the Haversian canals, and the interior of the hollow bones at all points that present augmentation by growth. They are therefore not peculiar to the periosteum. In the case of old persons and animals, when the growth of the bones has ceased, the osteoblasts are lessened in numbers, and have changed their forms in such manner as to lie upon the surface of the bone as thin flattened scales, which often can not be seen upon the margin in sections cut perpendicular to the surface; but in such sections they will appear whenever a Haversian canal is so cut as to present the flat sides of the cells to view, especially if stained with a good nucleus tinting dye. In this condition the cells seem to be inactive. In the study of young bones many regions of inactivity may be met with in which the osteoblasts present this appearance.

The function of the osteoblasts is clearly the formation of bone. There is no growth of bone without their presence. It is true that calcifications of tissue occur in various places without the presence of osteoblasts, and to the naked eye these may closely resemble bone; but upon microscopic examination they are found not to present the tissue forms of bone. These tissue forms are directly the product of the osteoblasts. The precise manner of the formation of bone is not agreed upon, two theories still being entertained. These may be briefly stated. The one view regards the osteoblasts as forming the matrix by aggregating themselves together upon the surface of the bone. This matrix thus formed is in turn converted into bone by becoming infiltrated with lime salts. All bone is shown by certain processes of chemical solution, to be composed of delicate laminæ laid the one upon the other horizontal to the growing surface, whether this be the surface of the bone proper, or the surface of the Haversian canals. It is supposed, according to this view, that these laminæ are made up from the different layers of consolidated osteoblasts. In this process certain of the osteoblasts persist, or are included in the formed bone without calcification, and thus become the bone corpuscles.

There are many objections to this view. One of the most potent is the fact that the osteoblastic layer is very rarely found in a condition of even semi-consolidation. The cells do not approach the loss of individuality necessary to the formation of a continuous sheet of matrix.

In reasonably good preparations they always appear as individualized cells. Furthermore, we are not able by any treatment of bone yet devised, to render the outlines of such cellular elements of its matrix apparent. These objections to this view has been pointed out by a number of prominent histologists.

Another view is that the osteoblasts shed out from themselves the material that forms the bone by some process closely akin to secretion, if it be not this in fact. It seems probable that this process of secretion is per-formed by all of these cells that lie against the bone, and that the process is not continuous, but presents alternations of activity and rest. This will account for the lamellation observed in bone more perfectly perhaps, than the supposition previously mentioned. This kind of lamel-lation is also observed in the structure of the shells of shellfish, the formation of which is generally agreed to be by a process of secretion. In the formation of bone by this process, certain cells seem to become matured and flattened down against the surface, and to sink beneath it. As a matter of fact, the bone material is built up over them and they become encapsuled, and are then known as bone corpuscles. They lose bulk in this process, so that the bone corpuscle is usually smaller than the original osteoblast. It is difficult to see the processes of the bone corpuscles in moist specimens, but they are plainly apparent in sections of dried bone, in which the canaliculi, which were occupied by them, are filled with air. The cells that sink into the bone in this manner, while not entirely regular in number and distance from each other, do present a kind of regularity which serves to give the impression of rows around the Haversian canals, and along the borders of sub-periosteal bone (in cross sections of the long bones). These rows bear a pretty distinct relation to the lamellæ of the bone, as would naturally be expected if either of these explana-

6

tions of the process be adopted. The osteoblasts in flattening down very generally lie lengthwise upon the long bones, therefore the resulting bone corpuscle lies in the same manner with its broadest diameter to the forming surface, whether this surface be that of the wall of a Haversian canal, or the surface of the bone. Therefore, in cross sections of the long bones we get cross sections of these cells, so that they present a somewhat different appearance from that seen in the lengthwise sections.

The processes of the bone corpuscles are very numerous, and radiate in every direction through the bone matrix forming junctions with each other. (See fig. 25 c.) Each individual bone corpuscle with its processes seems to preside over a specific area of bone matrix, and the impression might be entertained that this individual corpuscle had formed this area. This impression is also much strengthened when in the study of irregular formations, globules of bone are found, each showing a single bone corpuscle near its center; or perhaps several of these lying together with the area of each more or less clearly visible. In studying these, it is often difficult to escape the conviction that each osteoblast that so matures as to become a bone corpuscle really forms the area of bone with which it is immediately surrounded. However, the study of the forming surface will serve to dispel this idea and admit the assistance of osteoblasts not yet so fully matured. Again, a close study of the processes of the bone corpuscles shows that their general direction is perpendicular to the forming surface of the bone (fig. 25), so much so that with low powers a striation in this direction often becomes prominent. Much of this is due to the processes reaching far into the bone before the encapsuling occurs.

In those portions of bone formed under an attached periosteum, particularly if the penetrating fibers are numerous and large as shown in figs. 20, 21, 23 and 24 the bone corpuscles do not lie with their long axes horizontal to the surface of the bone, but in a line parallel with the penetrating or residual fibers. This is explained by the fact that the osteoblasts are held, or lie between the fibers in such a way as to present their short diameters or ends to the surface of the bone, which position they retain. It serves as a mark designating the portions of the bone formed under a periosteum of this character even when the fibers themselves can not be seen. This applies to all those forms of the attached periosteum in which the penetrating fibers are large and thickly set. When the fibers are more sparsely distributed the osteoblasts and the resulting bone corpuscles, may lie in the same relative position to the forming surface, as in case of the non-attached forms.

The osteoclasts, myoplaxes or giant-cells, present various forms, vary indefinitely in size, and are usually multinucleated. (See figs. 19 f, f, f, 24 g and 27, a, a.) Occasionally, one may be recognized with but a single nucleus, and I have seen them containing as many as twenty-four. From four to ten is a more common number. The general inclination is to the round or oblong form. They are very rarely branched and present no processes. Such forms of cell may be found in other localities, and we can only recognize them definitely as osteoclasts, when found in contact with bone, or some of the hard tissues undergoing absorption. In such positions, they uniformly lie in little bay-like excavations in the surface, known as the lacunæ of Howship. They conform in certain measure to the depth and size of the excavation in which they lie, which fact seems to argue that most of their growth has occurred in this position. I often see very small ones in small excavations and large ones in correspondingly large excavations. But in absorption of greater extent, such as in the hollowing out of the shafts of the long bones, we often find very large cells

lying on the surface of the bone without any lacunæ whatever. I may say, however, that the number of such cells
that are sometimes seen in the tissues of the bone marrow,
detached from the bone, but in the neighborhood of
extensive absorption, has given me the impression that
possibly these cells may in some degree possess amœboid
movement during life, and therefore, a limited power of
migration.

The function of these cells is sufficiently obvious.
They dissolve the bone with which they are in contact,
probably by the secretion of a solvent fluid, making room
for themselves, and in this way remove the surface of the
bone, i. e., cause its absorption. In this way the enlarged ends of the bone are trimmed down to the size of
the shaft (in the elongation of the bones during growth),
and the central cavities are hollowed out. Channels are
burrowed through and new bone again deposited, thus removing the old and filling in with new. In this work the
osteoblasts and the osteoclasts are continually replacing
each other, the osteoblasts building and the osteoclasts
tearing down ; and by the joint action of these, both the
formation and the conformation of the bones are affected.

CHAPTER VI.

Histologists have usually described three modes of the formation of bone. These relate to the conditions under which the bone is formed, and are the subperiosteal, intra-cartilaginous and intra-membranous. The subperiosteal is not a *de novo* origin, but a growth superadded to previously formed bone, or laid down upon the surface of cartilage. This has been in a measure considered while describing the functions of the osteoblasts, but certain points should be elucidated. The surface of a growing bone is not smooth and compact, but is continually thrown into convolutions by the upward growth of spiculæ, or upon long bones, of long ridges more or less sharp, arranged parallel with the long axis and which often rising into the tissues of the periosteum, form arches by spreading laterally and joining with like ridges on either side, as shown in figs. 23 *a, a,* and 26 *c, c, c,* both of which are from sections cut across the long axis of the formation. In this manner new Haversian canals are being formed into which capillary vessels are sent to supply the parts with blood. By this means the extent of surface to which the osteoblasts are applied for the building of bone is immensely augmented. As one set of Haversian canals is completed in this manner, indeed often before their completion, new spiculæ, or ridges of bone, are again projected from the surface to form others exterior to these. The Haversian canals thus formed are already lined with osteoblasts, which continue the work of deposit of bone upon their walls, and thus the bone grows not only upon its surface, but on the walls of the canals also. This

45

is best seen in fig. 26, in which a comparatively low power
is used for the illustration of a portion of a cross section
from the tibia of the kitten.   At c, c, are shown upward
growths of spiculæ, into the tissue of the inner layer of
the periosteum.   At d, these are throwing out processes
from their summits.   At e, these are just meeting together, and at f, fully formed canals are seen.   It will be
noted that the bone is thickly studded with them.

The soft tissue as it is included in these Haversian
canals changes its type very noticeably, losing most of its
fibers, and becoming still more like fetal tissue, presenting
many undeveloped cells which in the main lie quite widely
apart, giving a tissue of very simple character.

As has been indicated, the subperiosteal bone is
deposited in laminæ that are concentric to the axis of the
shaft of the bone, and the position of the bone corpuscles
corresponds with these, presenting their flattened sides
to the surface, and therefore in the absence of these convolutions of the surface, present laminæ arranged around
the shaft.   But in this convolution of the surface, the
arrangement of these laminæ, and also the position of the
bone corpuscles is in accord with the surface at the time
of its formation.   It will be seen that this produces a
seeming confusion in the laminæ and position of the
corpuscles in the portions of bone thus formed.   The
bone formed on the walls of a Haversian canal presents
concentric rings around the axis of the canal to which the
bone corpuscles also present their flattened sides, thus distinguishing this bone from the subperiosteal formation.
These rings of Haversian bone are known as Haversian
systems, and they play a very important part in the
growth of bone.

This subperiosteal formation of Haversian canals is not
confined to the non-attached forms of the periosteum, but
occurs beneath the attached also, though less regularly.
It seems that where the penetrating fibers are very large

and numerous, very few such canals are formed. In these the upward growth of the spiculæ is generally along the line of some of the fibers, and as the arches are thrown out on either side, the fibers of the region are included in the bony formations without disturbing their position, but that portion of the fibers which is included in the Haversian canal soon disappears, so that the Haversian bone has very few or none of these. (Figs. 21 *h, h,* and 23 *b, b, b, b.*) In some instances in which the fibers are unusually large and strong, as in the alveolar process, which I will describe in connection with the peridental membrane, the fibers are often seen stretching across fully formed Haversian canals, thus showing a tendency to persist, or that large and strong fibers are removed more slowly. In many instances, however, especially about the bones of the face where there are unusually strong periosteal attachments, the sub-periosteal growth of bone is originally destitute of Haversian canals and thickly set with penetrating, or residual fibers. These fibers are not a part of the bone *per se*, but are the accident of the formation. In other words, the method of making firm hold upon the bone, is the implantation of the fibers by building the bone about them; and after enough of the length has been included in this way to serve that purpose, the deeper portions are of no further use, and as a matter of fact, are, according to my own observation, removed.

This is accomplished by the removal, not of the fibers simply, but of the whole mass of bone thus formed, by the successive burrowing of Haversian canals through its substance, and the formation of what may very properly be termed secondary Haversian systems. I have illustrated this in fig. 24 from a section from the lower jaw under a muscular attachment. Here it will be seen that the penetrating fibers of the periosteum are strong and thickly set, and though the surface of the bone presents some low spiculæ reaching out along the line of some of the fibers it is really a solid growth of subperiosteal bone. A little way inward from the surface, however, there is a Haversian canal which fortunately presents a direct lengthwise section. On the walls of this canal from the letter *c* to the line drawn across at *e*, Haversian bone is deposited, and the canal itself is lined with osteoblasts. Above the line *e*, there is no deposit of Haversian bone, and instead of osteoblasts, the walls are lined with osteoclasts, which lie in little bay like excavations, *g*. In this part of the canal the excavation of the subperiosteal bone is in active progress, while in the parts below the line *e*, the excavation is being filled up with Haversian bone. This is readily distinguished from the subperiosteal bone by noticing three points. 1st. The Haversian bone is slightly different in shade from the subperiosteal. 2d. The long axis of the bone corpuscles lie in a different position, or in the direction of the long axis of the Haversian canal. 3d. It has no residual fibers. Now if we scrutinize the original walls of the canal where the Haversian bone joins the subperiosteal, it will be seen that they are everywhere indented with the bay like excavations as pointed out at *f*, *f*, *f*. These may be contrasted with similar Haversian growths, shown in Haversian canals of subperiosteal formation, in figs. 21 and 25, which do not show these bay like forms. Wherever these are found they are the marks of the work of the osteoclasts, and show the canals, or Haversian systems, deposited therein to be secondary formations, that is to say, formed by the removal of previously formed bone. In my studies of the bones of the jaws in animals of different ages, I find that almost the entire original formation of bone is removed in this way, and replaced by Haversian bone. This seems to be especially the case at all points where there are many residual fibers, so that in old bones these fibers penetrate to but a slight depth, otherwise they occur only at isolated spots that have been missed by the process of removal. In the growth of the long bones, where the residual fibers are localized at particular points, the same removal seems to occur. In the regions of the non-attached periosteum this kind of regeneration may also be found, but it seems not so complete.

These residual fibers are known in our literature as penetrating fibers, or fibers of Sharpey, who first described them as fibers penetrating the laminæ, i. e., passing in a direction transverse to the laminæ of bone. While this observer with others who have followed him, seems to have recognized that these were in many instances derived from the periosteum, and were irregular in their occurrence, it seems that the fundamental reason for their appearance was missed by them, which is probably to be explained by the fact that in their study of the subject, the variety of the fibrous forms and purposes of the periosteum had not attracted special attention. It will be clearly seen that it is only by a close study of this membrane in different positions, giving attention to the purposes it subserves, and the forms adapted to these, that we may gain a clue to the uses of these fibers, and become able to know in advance where to find them in abundance and thus be able to follow up the study of them satisfactorily. These localities have been sufficiently indicated.

The function of these fibers, it seems to me, is physical and entirely passive; that of giving firm attachment to the periosteum and the tissues it supports. They may give direction in many instances to the growth of bone by forming a kind of ladder, or central line, around which osteoblasts may concentrate, but in no other way contributing to that growth. The active functioning process is to be sought for in the act of the formation of the fiber, not in the fiber after its formation, and I see no reason for considering these in any other light than as white connective tissue fibers, subserving a specific purpose, that of giving support. I should, therefore, regard the term

7

osteogenetic fibers consisting of *osteogenetic substance*, applied to them by Sharpey and adopted by many histologists, as an error. It is true, as pointed out by the above
named observer and others, that in the upward growth of
spiculæ of bone, for the formation of Haversian canals,
the line of these fibers, if they are present as.penetrating
fibers, is followed, and that the osteoblasts are arranged
about them and seem to be clinging to them. It is also
true, that we find these osteoblasts arranged about a center of advance of bony spiculæ in the absence of such
fibers, where the growth is taking place under the non-
attached forms of the periosteum. I have scrutinized this
point closely, employing as aids the various plans of
preparation and staining, without being able to make out
any such fiber or fibers, that would seem to be in any way
directing the growth, but on the contrary, have found the
growth of such spiculæ proceeding across the line of the
horizontal fibers of the non-attached forms of the periosteum. Again, as has already been stated, even in the attached forms, where the upward growth of spiculæ follows
the line of particular fibers, the processes given off at
either side, forming the arch connecting with neighboring
spiculæ, for the formation of Haversian canals, pass
directly across the direction of these fibers without disturbance of their position. These facts show that the
formed fibers become passive physical agents, and have
not a genetic function. It is probable that in many
instances, as in the intra-membranous formation of bone,
which will be studied later, the formation of fibers is a
necessary step in advance, affording a kind of framework
or lattice, as a basis for the new growth, but in every instance, the work of deposit of new bone is performed in
the presence of, and by, the osteoblasts, which are specialized from the embryonic connective tissue cells, and are in
no way dependent upon the fibers, except as these afford

a meshwork, in which they may undergo their developmental stage.

The condition of these fibers as to calcification is of
interest, although not very positively made out. In
many instances they appear to have become stiffened by
the reception of lime salts in advance of the line of the
forming bone, and thus become a nidus for the upward
growth of spiculæ. In other instances appearances indicate quite the reverse condition, and in some very thin
sections cut across the fibers, I have found them so loosely
attached near the border of the forming bone, that they
have fallen out of their alveoli. Occasionally quite deep
in the substance of the bone I have found them protruding
from the broken margins of the section, as illustrated at
g, *g*, *g*, in fig. 21. Furthermore, in pulling apart the laminæ of subperiosteal bone, we may find them withdrawn
to some length as has been so admirably illustrated by
Sharpey. (Quain's Anatomy.) These facts show the
loose connection of the fibers with the bone substance,
but it must be remembered that this is only after the with-
drawal of the lime salts by the process of decalcification
in the preparation of the tissue. Before the decalcification, these points can not be demonstrated. It therefore
appears that the connection of these fibers is not intimate
with the basis substance of the bone, but that they are
rendered firm and seemingly of equal consistence with
that matrix, by the common reception of lime salts, in the
process of calcification.

With the description given above, and the illustrations
presented, it will become clear that isolated patches of
these fibers may be found that do not reach the surface of
the bone, having been cut off by the formation of Haversian systems. Again the fibers may be cut away by
absorptions occurring beneath the periosteum, the portions so absorbed being refilled with bone, and the subperiosteal growth proceeding as before. I have been so

fortunate as to meet with examples of this in my sections, one of which is presented in fig. 27. This fact serves also to explain some things which heretofore seemed dark in the formation of the cementum on the roots of the teeth, which will be studied later. In the figure it will be seen that at *a, a,* there are several osteoclasts lying in deep excavations, and at *b, b,* there are points at the unabsorbed surface of the bone in which the fibers of the periosteum are seen implanted. The fibers appear also deep in the bone, and it will be seen that in the region of the absorption *c, c,* these have been removed, together with the bone, and the space filled with tissue of a marked fetal type, but containing a large number of developing cells. In many instances I have seen such absorptions that have been refilled with bone, and the fibers reattached only after much of this secondary bone had been deposited. This figure is taken from the margin of a considerable absorption which was taking place on the inner (lingual) surface of the lower jaw, about the position of the cuspid tooth and near the attachment of the mylo-hyoid muscle, and was probably affecting some change in the form of the bone. It shows the nature of the absorptive process. Formerly it was supposed that the bone corpuscles took part in the process, enlarging the capsule in which they are lodged, but my own observation shows quite conclusively that they take no part whatever. It is the work of the osteoclasts entirely, at least in physiological absorptions.

Intra-membranous formation of bone occurs only in the tabular bones of the cranium and face, and possibly a portion of the clavicle. Its only difference from the subperiosteal formation is that the bone arises *de novo* in membrane. In all other respects it is subperiosteal. The description of it may therefore be limited to the beginnings of the formation. The form is first laid down in what appears to be ordinary fibrous membrane, in which there is seen a tendency of the fibers to aggregate into bundles, which are often condensed into fibers so large as to almost merit the name coarse fibers. The bulk of the membrane is, however, composed of fine white connective tissue fibers. These decussate with considerable freedom, seemingly with a tendency to form an irregular meshwork. The fibroblasts are abundant and of the usual form, and there is seen quite a large number of undeveloped connective tissue cells. At the point where bone is about to be laid down, these latter may be noticed to aggregate themselves together and grow larger, elongating in a direction parallel to the fibers with which they are associated, but not becoming distinctly fusiform. These come to be closely packed together at a single point, and they may form a row of some length, or elongated islands, which take staining agents more strongly than other tissues. Passing these in a direction toward the more central parts where bone formation has begun, it will be found that this is laid down in the center of an exactly similar cluster of cells. As the deposit of matrix and lime salts is laid down, the cells seem to spread asunder as if to make room, while one, two or more remain in the calcified mass as bone corpuscles. These islands may be found of any dimensions, from that of the smallest island in which a single cell may be distinctly recognized as a bone corpuscle, to considerable areas of bone, all presenting the same characters. The usual forms of the osteoblasts are seen on the margins of the formed bone. I have represented one of these islands in fig. 28, from the parietal bone of the human fetus in which b, b, point out individual osteoblasts and a, a, bone corpuscles. The latter appear unusually large in all of the very young intramembranous bone that I have examined. The osteoblasts cluster very thickly around the margins. The fibers of the membrane do not at first give place to the bone formation, but are included within it, as shown in the figure,

and those osteoblasts that lie at the ends of the formation
are apt to present their ends to it, seemingly constrained
to this position by the presence of the fibers, as ·in the
attached periosteum. Different islands present much difference as to the included fibers, some having very few,
while others, as the one chosen for illustration, have
many and the effect of these on the disposition of the
cells around the margins is plainly apparent. For a little
way outside the layer of osteoblasts that are in contact
with the formed bone, the developing cells are thickly
placed, and are evidently destined to become osteoblasts.
The islands of bone thus formed grow into spiculæ and in
time unite with others in the neighborhood, and at first
form a kind of bony latticework, the openings in which
are finally filled to form the complete bone. This is then
extended until joined to its neighbors by suture.

CHAPTER VII.

GROWTH OF BONE UNDER TENDINOUS ATTACHMENTS—
INTRA-CARTILAGINOUS FORMATION OF BONE.

As we have seen, in the more rigidly attached forms of the periosteum, where the fibers are very thickly set, the growth of bone is so modified that there is no formation of Haversian canals by the growth of spiculæ and arching over, but the bone is deposited in a solid surface and afterwards burrowed out for the formation of these canals. This kind of fibrous investment of the surface of the bone is the limit at which osteoblasts appear on the growing surface. At the points of attachment of tendons the strong fibrous bursæ, no osteoblasts appear on the surface of the bone, and its extension by growth under these conditions is effected upon a different plan. This is accomplished by the projection of Haversian canals, from the bone beneath into the tendon, dissolving a portion of the substance of this, and depositing Haversian bone on the walls of these canals.

I present in fig. 29 an illustration of this process as seen at the attachment of Tendo Achillis. The section is cut from a young lamb lengthwise of the tendon fibers. The large fibers of the tendon are represented at *A*. For a little space in advance of the bony formation these are partially converted into fibro-cartilage, and the cartilage cells are shown mostly between them. The fibers of the tendon are, however, easily separated with needles down very close to the new bone. Indeed, although the cells appear as shown in the illustration, the tendon does not otherwise appear to be cartilaginous.

At *B*, Haversian canals, or at least capillary loops, grow out from the bone beneath and penetrate the fibers,

removing their substance.   A number of these are seen in
the figure.   This growth presents absorption-cells similar
to those met with in the absorption of bone, but they do
not seem to attain so large a size, and react differently to
staining agents, which seems to indicate that there is
some chemical difference between them.   Indeed, I have
found some difficulty in so staining these cells as to differentiate them clearly.   They may be made out, however,
by high powers, if the sections are sufficiently thin, without staining.   With these absorption-cells many embryonal
elements are associated, in such a manner as to very
effectually obscure them if there is much thickness.   After
this process of absorption has proceeded for a certain
space, in advance of the surface of the bone, the processes going on within the new canals are changed.
Osteoblasts take the place of the absorption-cells, and
bone is laid down upon the walls of the excavation.   This
is shown at D, D, while c, c, c, show the Haversian canals.
These are older portions of the Haversian canals, and it
will be noted that the bone is formed upon or laid against
the tendon fibers, which here are more or less calcified, or
infiltrated with lime salts.   Everywhere along the sides
of these formations of bone will be seen the bay-like
forms due to the absorbent-cells.   The growth is composed entirely of Haversian bone.   The penetration of
the tendon by the individual canals is not always in the
line of its fibers.   Though in a general sense it is so, individual canals are often seen to diverge very considerably
from this line.   In the examination of sections *in extenso*,
it is found that these canals branch in various directions
into the tendon, and that new canals are formed very frequently by the absorbents piercing the sides of the
Haversian systems formed, and starting out in new directions.

It will be noticed that there is always a very large
portion of the fibers of the tendon left attached to the

formed bone. Were it otherwise, the strength of the attachment of the tendon might be seriously impaired. Really many of the fibers continue to penetrate deeply among the Haversian systems, as seen at *E*. These are finally removed by the formation of new Haversian canals, by which the new bone soon becomes very much cancellated, though we may find occasional isolated patches deep in the formed bone.

The conversion of the articular cartilages into bone, during the lengthening of the shaft, is a process almost precisely similar to that just described, so far as the absorption and removal of the cartilage, and the deposit of the new bone is concerned. There is, however, a marked change in the cartilage, which appears to be a preparation of it for removal and the formation of bone. I should state here, however, that there are two quite distinct modes in the replacement of cartilage by bone, the one taking place in the articular cartilages of the long bones, and probably in the whole of the short bones, while the other is confined to the shafts of the tubular bones. I do not wish to confound these two forms of the process, as has so frequently been done in our past literature.   .

I present an illustration from the tibia of a kitten, representing the position of these two forms soon after the process in the epiphysis, fig. 32. The process of ossification has begun in the diaphysis early in fetal life, and is far advanced before the beginning is made in the epiphysis. The subperiosteal growth of bone is begun before the beginning of the intra-cartilaginous, and continually precedes it. *c, c,* Represent the subperiosteal deposit, the advance of which terminates in the periosteal notch *e, e.* It will be noted that a thin layer of subperiosteal bone passes up to this notch. *f,* Marks the beginning of change in the cartilage cells represented in fig. 33, and *g,* the line of the absorption of the cartilage and

8

the point of the diaphysal intra-cartilaginous formation of bone. The darkened portions d, represent the bone marrow, and the light portions the bone formed, which latter occupies but a small part of the area. At a, the cartilaginous head of the bone is seen, with the beginning of the process of epiphysal ossification at h.

The changes in the epiphysal cartilages will appear somewhat differently in the same animal at different ages. In following up the examination of sections, cut through the center of the head of the tibia, for instance, beginning with the surface of the articular cartilage, this will be found to consist almost entirely of fine white fibers (fibrous tissue), arranged perpendicular to the surface, and lying very compactly together. A little inward from the surface very small cells come into view, and if the section be sufficiently thin it will be noted that these lie between the fibers. Still proceeding inward, the cells become larger, and the fibers more indistinct. In many instances the fibers seem to gather into groups or bundles, and terminate; in others, they simply fade away into the cartilage. Farther inward, nothing but the clear ground substance of the cartilage studded with its cells, is seen. These cells are small at first and single, but they grow larger, and soon we find two together, and finally three or four, and perhaps more, which seem to occupy a single capsule.

At this point a great variety of appearances may be found in the examination of the changes as they occur in animals at different ages. In fig. 31, I present an illustration from the head of the tibia of a lamb about two months old, which shows these cells as flattened and in distinct rows. In this case the change of the cartilage to bone is well advanced. If it had been examined at an earlier age, especially before the junction of the diaphysal with the epiphysal bone formation, these rows of cells would not have been found, but instead the enlargement

of the cells would have been found proceeding regularly,
while they remain scattered without order, i. e., without
falling into rows. Although I have not seen the actual
division of cells, and know of no one who claims to have
done so, there can be no reasonable doubt that we have
been following the active results of growth, not only of
the individual cells, but of the multiplication of the cellular elements by division. After passing inward from
this formation of rows, it is found that this arrangement
is lost, and the cells are again scattered over the field
without order of arrangement, as shown in the figure, and
that the space between them has become much greater.
At this point the cartilage becomes infiltrated with lime
salts, and all activity within it ceases. This statement
should, however, be qualified, in so far as to say that in
the younger articular cartilages, when the process of
change is just beginning, only the matrix becomes calcified, the cells remaining soft; but in such as that from
which our illustration was taken, the cell bodies are also
calcified.

The process of absorption of the cartilage is begun by
the perforation to its center of one or more canals. These
then spread out for a considerable space, showing at first
no disposition to the formation of bone. The cartilage is
absorbed, and replaced by tissue of a primitive fetal type.
Soon, however, short canals begin to radiate from this
central cavity, and after proceeding for a space, the process of absorption ceases. Osteoblasts then develop along
the walls of the space opened, and begin to lay down bone
upon the cartilage. While this is in progress, new canals
are being opened at other points, which in succession
receive deposits of bone upon their walls. In this process the new bone is laid directly upon the cartilage, and
in the successive burrowings, very nearly all of the cartilage is removed, but often a small remainder will be found
at some distance from the margin where the absorption is

in progress, in the central part, perhaps, of some portion of the newly formed bone. However, the first bone laid down is usually removed by absorption after a time, so that, in the end, all of the cartilage is absorbed. Indeed, at first—that is, in very young animals—when the absorption of the cartilage is still confined to its central parts, the Haversian canals formed remain very large, only a thin stratum of bone being deposited upon their walls. At a more advanced age, the deposit of bone is greater, though at all times it remains quite cancellous.

During the earlier period of absorption the cells of the cartilage remain soft, and the effect of this condition influences the absorptive process, for as soon as the wall of one of these is opened the cell body seems to escape, at least it disappears, and the place is occupied by fetal tissue of the same type as that which fills the Haversian canals in the immediate neighborhood. There has been much speculation as to the destiny of the cartilage cells, some supposing that they form osteoblasts by a process of division, but it seems now so well settled that they simply disintegrate before the absorbents that it is hardly necessary to discuss the point. I may say that I, like many others who have essayed to examine this point, have been unable to see just what does become of them. They seem simply to disappear.

In older animals, when this portion of the cartilage in its entirety, including the cells, becomes calcified the process of absorption is studied to much better advantage; for before this time, the walls upon which the absorbents act are continually being broken by the opening of the capsules, with the protrusion of the fetal cells of the Haversian canals into them, and these are so numerous as to obscure other phenomena. But now the absorption of the calcified cells proceeds in the same orderly way as in the matrix itself, and the wall acted upon by the absorbent cells is continuous around the new Haversian canal, as shown in the illustration. Here we may trace the work of the chondroclasts, both by the presence of these cells, and the indentations, or bay-like cavities left in the walls of the cartilage, both in the new Haversian canals, in which bone is not yet deposited, and in those in which it is, as is shown in the figure. In order to illustrate this to better advantage, I have prepared a representation of a single one of these absorbed spaces, or new Haversian canals in fig 30, using for the purpose a high power, in which the chondroclasts, and their relation to the liquefying cartilage can be studied to better advantage. These are identical in form and function with the osteoclasts, and here should be termed chondroclasts, to indicate their position, as on the same principle we call them osteoclasts when found applied to the solution of bone. Here we find them applied in precisely the same way, and producing similar results. Touching the question of the destiny of the cartilage-cells, I may say that under the conditions last described, I have often found one-half of one of these cut away by the absorbents, while the other remained in its matrix, which, it seems to me, definitely settles the question of the destiny of these cells, when in this condition at least. They are simply absorbed in the same way as the cartilage matrix.

The regular placement of the chondroclasts, in the absorption of the articular cartilage, can not well be made out before the calcification of the cartilage-cells. Their presence, however, is sufficiently manifest. They may be found about the walls of the liquefying cartilage, at many points, soon after the beginning of the process, but there seems to be no regularity whatever in their distribution, so that in studying the process in this stage alone, one could not easily make out that these were the principal agents of solution. Indeed, it is extremely doubtful whether they are the only agents of this process, as we shall see in the study of that form which takes place in the shafts of tubular bones.

The articular cartilages are continuously increased by growth and division or multiplication of their cells, and as continuously absorbed and replaced by bone until the process ceases with the cessation of growth at maturity. This growth of the cartilage represents the lengthening of the shafts of the bones. This lengthening takes place mostly in the diaphysal form of the process, until that ceases by junction with the epiphysal, and it is at this time that we find the cartilage-cells of the epiphysis falling into rows before the advance of the absorbents. It seems therefore that the lengthening of the bones takes place mostly by the growth of the cartilage in this particular form, which will be described more particularly later.

As has been said, the greater part of the skeleton is first laid down in cartilage. This is true of all the long bones, and at a later period this provisional cartilage is replaced by bone. The mode by which this change is made in the diaphysis of the tubular bones, is distinctly different from that in the epiphysis. This relates especially to that portion, which is at a later period converted into the bone marrow, which really, at this time, includes the whole of the shaft of the bone. The beginning of this process is seen first, near the middle of the length, and in the central part of the cartilage representing the bone. Here the cartilage cells are seen to enlarge, and apparently move apart, so that in addition to the enlargement of the cells, there is also an augmentation of the matrix as well. Just how the increase of this matrix occurs is not well understood. Indeed, I may say that although the processes taking place in the growth of the cartilage have long since strongly attracted the attention of histologists, they have not yet been made out satisfactorily. The fact of growth in this portion, and decided increased

activity over that of the other parts of the cartilage is
stfficiently apparent in the facts just stated, and, at the
same time, there is a slight but decided enlargement of
the diameter at this point, though the principal growth is
in the direction of the length of the shaft. At this time
a change becomes manifest upon the surface of the enlarged portion. The internal portion of the perichondrium is composed of moderately fine white fibers running
for the most part lengthwise of the cartilage, and lying
quite compactly together. Between these the cartilage
cells arise; first as very fine white granular elongated
points not unlike the fusiform cells of white fibrous
tissue. Successively they grow larger, as we proceed inward from the surface, and finally the fibers fade away
into the fully formed cartilage, without any very strict
limiting line. Occasionally however, the change to cartilage is more abrupt, and there seems to be a more or less
distinct limiting line between the perichondrium and the
cartilage. Especially will this be the case in a cross section, or if the section is somewhat diagonal. Within the
borders of the growing cartilage, the cells are still very
nearly of the same form, but larger, and as we proceed
inward, they gradually assume the usual form of the cartilage cells. Now at the point, where the cartilage cells
have enlarged in the central part, and the enlargement of
the diameter of the cartilage is seen, a change takes place
in the inner layers of the perichondrium. Fetal cells,
small round or oval cells, are deposited here and gradually separate the cartilage and perichondrium. This portion assumes a distinctly cellular type, and if the processes are carefully followed, with high powers, in sufficiently thin sections, it will become evident that some of
these cells become fusiform (fibroblasts) and are proceeding to the development of white fibrous tissue, while
others grow larger, and become arranged along the margin of the cartilage. The appearance of a layer of bone

deposited on the surface of the cartilage soon distinguishes these as osteoblasts.   Here it is perfectly evident
that an inner layer of periosteum has had a *de novo* development.   This portion is evidently not a change of
the perichondrium into periosteum.   The change in the
outer layer of coarse fibers is not so clear, for though
some change is gradually taking place in the form and
disposition of the fibers, it does not appear to be the sub-
stitution of a newly developed tissue, as is the case in the
inner layer.   Such is the beginning of the deposit of subperiosteal bone upon the cartilaginous matrix, and this
goes on coincidentally with the changes occurring within
the central portions; for while the changes just described
are in progress several capillaries have pierced the forming layer of bone, and found their way, by a process of
absorption, to the very center of the cartilage, where the
cells are most enlarged.   Here the capillary loops enlarge, melting down the cartilage about them and forming
a central cavity.   This is filled with tissue of the fetal
type, in which very small round cells (marrow cells) are
especially abundant.   These are evidently leucocytes deposited here from the capillary vessels.   While these processes are in progress, further changes have occurred in
the cartilage itself.   The cartilage cells throughout its
thickness become enlarged.   This change spreads either
way toward the extremities for space, and these enlarged
cells are seen to have fallen somewhat irregularly into
rows, which coincide with the long axis of the cartilage.
Very soon, or, we may say, coincidentally with these
changes, areas of proliferation of cartilage cells are established on either side (toward either end of the future
bone) of these enlarged cells, which are maintained
throughout the period of the removal of the cartilage, and
its replacement by bone.   In fig. 33 I present an illustration from a pigmented section cut from the femur of a
kitten, of the changes taking place toward one end of the

future bone, after considerable progress has been made in the process. At *A*, the unchanged cartilage is seen, and the progressive changes are represented between that and the letter *F*, which latter shows a portion of the bone-forming area. Between *A* and *B* the cartilage cells suddenly become smaller, and the supposition is that this is by subdivision. In this process they fall into rows, which is continued to *E*. It will be noted that very soon the cells grow larger, and continue to do so, down as far as the letter *D*. In the study of complete sections, it is found that this growth is mostly in the direction of the length of the bone. Indeed, this is very apparent in the illustration, for, it will be noted, that at *B*, the cells are very much flattened, so that in the lengthwise section they appear banana-shaped with rather a disposition to enlarge at one end ; and, in the enlargement from *B* to *C*, they become somewhat rounded, mostly by gaining in breadth. At *C* also the capsules that enclose the individual cells are more and more separated in the direction of the forming bone. This represents the growth of the shaft of the bone in length, for after the osseous substance is once formed, it does not increase in dimensions by interstitial growth. The increment is always to the ends, first in cartilage, then by the process of growth represented in the illustration. In the region of *D*, it will be seen that the nuclei (the transparent portion of the cell fills the capsule) are diminishing in size and becoming rather ragged in outline. This is undoubtedly a mark of degeneration in the cell, and it will be noted that the growth of the cells ceases from about this point. They become passive, and take no further active part in the processes that are going on ; unless it be a slight advance toward disintegration, as indicated by the progressive diminution of the nuclei. Occasionally a globular form will be seen occurring in the central part of the nucleus at this time, that closely resembles the marrow cells

9

which appear so abundantly after the capsules of the cartilage cells are broken into by the absorptive process ; but
I have never seen more than a single one of these in a
cell, and must suppose that this is either a nucleolus, or,
that it is an accidental dropping together of material of
the disintegrating nucleus.   In the region between D,
and E, that portion of the mass that forms the walls of
the capsules, in which the cells are imbedded, is undergoing the process of calcification, or infiltration with lime
salts.   At least, this is the general opinion of those who
have examined the subject with reference to this point.
My own examinations do not enable me to determine it.
The whole of the cartilage, from the region B down, is
much more transparent than the cartilage in which the
changes have not yet begun, and this difference is rendered much more prominent by pigmenting.   In this process the cell bodies remain absolutely transparent as
represented in the illustration.   This is entirely different
from the epiphysal cartilage while undergoing this
change.   In these the whole of the cartilate takes the
pigment in a marked degree, and this is true when the
two forms are included in the same section, as I have
often had them toward the end of the process of ossification.   They also react differently to other stains.   This
fact seems to show plainly that there is some chemical
difference in the structure, to which this difference of
reaction is due, but as yet I am ignorant as to what this
difference may be.

    The most interesting part of the process is that taking
place at E.   Here we find that the walls of the capsules,
facing toward the central part of the length of the shaft,
are successively disappearing before the advance of a
growth of fetal tissue, and that the remaining or lateral
walls of these same capsules become so many tubes in
which this growth advances.   Some of these, indeed, are
also broken down, merging two into one, frequently

making larger tubes. But for a space the greater number of them remain. This occurs, not in a few of the rows of cells that form the shaft of the cartilage, but in all of them together, generally forming very nearly a straight line through, or as seen in sections, across the shaft of the cartilage. That is to say, presenting a solid advance which includes the whole thickness as represented in Fig. 33. There are no radiating canals, as seen in the absorption of the epiphysal cartilages, or in the absorption of tendons, ligaments and bursæ. (Compare with Figs. 29 and 31.)

But before speaking further of the process of absorption, let us examine the tissue that is taking the place of the cartilage. At h, are seen the round cells (marrow cells) that are very generally advanced to fill the capsules from which the cells seem to escape, as soon as the capsules are opened. These are seen everywhere in the mass of fetal tissue, and they are often so abundant as to render the observation of the other cellular elements difficult. Furthermore, they are liable to be scattered over the other parts of the section, in the course of the preparation, and lead to confusion, unless special care is taken. On the right hand of the figure some of the tissue has been lost, but, in the other portions, it will be seen that these round cells are filling the capsules opened. At N, one seems to have been just opened, and these cells seem in the act of crowding into the aperture. At o, are pointed out fusiform or oval cells, applied closely to the remaining walls of the capsules which, it will be observed, extend to the lower end of the figure, and are marked with the letter p at various points. At m, m, m, m, are seen the cellular elements and fibers of the blood vessels, which are extended into each one of the opened tubes. j, Points out osteoblasts applied to the remaining walls of cartilaginous matrix, upon which no bone has yet been deposited. These are seen also in other parts of the

figure. No deposit of bone is seen usually, until the
space of several cartilage cells beyond the point of
absorption has been reached, but at *K, K, K, K*, is seen
deposited on the walls of the tubes a layer of bone which
is in turn covered with osteoblasts. The new bone is
deposited upon the remains of the cartilage, partially filling up the tubes formed by the opening of the capsules
of the cartilage cells.

We may now return to the absorption area, represented at *E*. In this absorption I have been unable to
identify a single well marked chondroclast, applied to the
opening of the capsules of the cartilage cells.   In Fig. 33
all of the capsules opened are partly destitute of cells,
i.e., have not yet filled up with the advancing cells, but in
Fig. 34 I supplement this defect, by choosing for illustration a few capsules which the fetal tissue fills compactly.
Here we find both fusiform and marrow cells applied to
the boneward face of the capsules, sometimes the one,
sometimes the other, or they may be mixed together.
These fusiform cells have uniformly a well marked nucleus of irregular outline, instead of the pale round form
seen in the chondroclasts and osteoclasts, and we see
apparently the same cells applied to the remaining walls
without witnessing any evidence of solution. The absorption cells appear in great abundance a little further away,
removing portions of the tube walls and enlarging the
ehannels.   (Fig. 35.)

The agents of solution in this case are probably the
small cells. It is well known that the leucocytes develop
this power in a marked degree in other localities. They
are the agents of solution of sponge, in the sponge-graft,
of animal membrane sutures, ligatures, and other foreign
substances. They are also the principal agents of absorption, when this process occurs in connection with inflammation. Furthermore, the absorption cells are undoubtedly developments from the primary connective tissue

cells, which these round cells represent. They require time to make their growth, and during this period of growth are exercising their peculiar function, as has been shown on a former page. It may, therefore, be assumed that in the absorption we are now considering, the amount of tissue removed at one point being only the thin wall of these capsules, is too small for the development of chondroclasts that will be capable of satisfactory differentiation from others by microscopic examination.

As already indicated, the remaining walls of the alveoli of the cartilage cells become the nidus for the deposit of bone, at a point removed, by the length of a few cells, from the point of absorption. It will be noted also that the greater portion of the matrix of the cartilage still remains, this having already been reduced to a very small amount, by the progressive enlargement of the cartilage cells, as illustrated in fig. 33. At a distance of a few cell-lengths further, the absorption of this remainder of the cartilage matrix, together with that of the greater part of the bone first deposited, is in active progress. This is best seen in cross sections, fig. 35. If these are double stained with hematoxylin and carmine, the remains of the cartilage will be purple, while the bone deposited upon it will be red, which distinguishes them sharply and quite beautifully, and at the same time the cellular elements are well shown. In following up serially cross-sections prepared in this way, and receding bone-ward from the point of the beginning of bone formation, it will be found that the Haversian canals are enlarged and decrease in number, and in this process all or nearly all of the remains of the cartilage is removed. Indeed, in the central part of the shaft only a little bone is formed, and this is all removed after a time, to form the cavity. Along the circumference the bony formation is stronger, and is merged with the subperiosteal bone; but

even this is also removed in time. In all of the central
portion of the length of the shaft it is not reformed,
or if so, it is only to be removed again ; but toward the
ends of the bone it is replaced with more mature Haversian bone.

The processes of bone formation and of its absorption
are going on simultaneously in a very close proximity. I
have illustrated this in fig. 35, from a cross section of the
rib of a young kitten, at some little distance boneward
from the point where ossification begins. *d*, Represents
a few Haversian canals cut across. *b, b,* Point out the
bone, and *a,* the remains of the cartilage matrix, which,
when there has been no absorption of the bone formed,
lies centrally in the irregular rings of bone. Osteoblasts
appear over a large part of the surface, but at some
points absorption is in progress; the osteoclasts are
indicated by the letter *c*. By the absorption overbal-
ancing the deposit of bone, the space is gained for the
bone marrow.

In all of these varying phases of bone formation, it
will be noted that the active agents are the osteoblasts.
These seem to be developed in the inner layer of the periosteum, or with equal facility in the tissue that fills the
Haversian canals, or the endosteum. They are therefore
not peculiar to the periosteum, but belong rather to the
surface of the bone, whether this surface be an external
or an internal one.

CHAPTER VIII.

The peridental membrane comprises that tissue which intervenes between the root of the tooth and the bony walls of its alveolus. It has received various names from time to time, as alveolo-dental membrane, dental periosteum, alveolo-dental periosteum, pericementum, etc. The office of this membrane may be regarded as threefold —functional, physical and sensory. It is functional in so far as it is the place of the development of the osteoblasts, which build portions of the alveolar walls, and the cementoblasts, which build the cementum. These cells seem to be received into the fibrous meshes of this membrane from the blood streams as leucocytes or amœboid cells and here undergo their development, or that differentiation which fits them for the building of bone on the one side, and the building of cementum on the other. During this development they become allied to their respective places, i. e., the surface of the bone and the surface of the cementum.

The physical function is the fixation of the tooth in its position, a passive function which is performed by the fibrous elements. These fibers, which I shall designate as the principal fibers, form the bulk of the tissue of the membrane, and their ends are fixed in the cementum of the tooth's root on the one side, and in the bone which forms the walls of the alveolus on the other, and are thus stretched across the intervening space in various directions, and in such a manner as to swing the tooth in its socket.

The sensory function is supplied by an abundance of nerves which enter the membrane from every direction through the walls of the alveolus, at the apical space, and by way of the gingival border below* the rim of the alveolus.

Besides the osteoblasts and cementoblasts, the membrane presents various cellular elements: such as fibroblasts for the augmentation or renewal of its fibrous tissues; osteoclasts for the removal of the walls, or portions of the walls of the alveolus for the accommodation of changes in the position of the tooth, or of the cementum for the change of the form of the tooth's root. These latter seem to be developed as occasion requires, but are very generally present somewhere within the alveolus. Besides the cells mentioned there is always a considerable number of undeveloped cells within the meshes of the fibrous tissue in young subjects, but not very many in the old. There is also a set of lymphatics which are peculiar to this membrane. They occur in great abundance immediately surrounding the cementum in young subjects, but are much diminished in numbers in the old.

In many parts of the membrane there is seen an indifferent inter-fibrous tissue. It is a tissue composed of cells and fibers, not possessing very marked characters, intervening between the principal fibers in many places, especially where these are large, and making up the bulk of the membrane in certain localities where these are absent, and accompanying the blood-vessels and nerves.

The form of the membrane is such as to closely invest the root of the tooth and fill its alveolus, but it does more than this, for it invests the tooth lower down than the lowest border of the alveolar wall. The membrane may conveniently be divided into three divisions: the apical, that portion surrounding the immediate apex of the root,

---

*In the descriptions which follow, the tooth will be regarded as a cone of which the end of the root is the apex and the crown, the base. Therefore toward the apex of the root is upward, and toward the crown is downward, whether the tooth is in the upper or lower jaw.

or occupying the apical space fig. 36 e ; the body of the membrane, which fills the alveolus from the apical space to the lower border or rim of the alveolar wall a ; and the cervical or gingival portion, or that portion immediately surrounding the neck of the tooth below the rim g.

The thickness of the membrane varies very much in different individuals, and in different teeth in the same individual. It is thickest in the child, and it becomes thinner (normally) as age advances until forty, or perhaps sixty years has been reached. In fig. 37, I give a very accurate outline of a cross section of the alveolus with its contents from a lamb (temporary tooth), cut at about the middle of the lower third of the body of the membrane ; and in fig. 38, another from the cuspid tooth of a man forty years old, cut a little closer to the gingival border, so as to include the thickened rim of the alveolar wall. These are drawn as enlarged by a two-inch lens (using a camera lucida), and then reduced one-half, and illustrate very fairly the extremes which occur, normally, in the thickness of this membrane. Such blood-vessels as could be clearly seen with this low power are shown in their proper positions and dimensions. It will be seen that these are usually midway between the alveolar wall and cementum, in the central part of the membrane, in the young subject ; while in the old they are more generally close to the bone, indeed very many of them lie in grooves in the alveolar wall.

The direction of the fibers of the membrane in the position of the sections is indicated as perfectly as is practicable with this low power of the microscope. Variations in the thickness of the membrane are presented in its different parts, but these seem to follow no rule whatever, except that it may be that the apical portion is generally somewhat thicker in young subjects. With this exception, the differences in thickness seem to be mere irregularities in the contour of the alveolus, which is constantly under-

10

going change by absorption and rebuilding of bone. The
general form of the membrane is better seen in fig. 36
from a lengthwise section of an incisor tooth of a young
kitten. This tooth, though very slender, is so small (only
three-sixteenths of an inch long) that it gives a better
opportunity for a full length illustration than the larger
teeth of man. The elements of the membrane are the
same, however, both in form and arrangement in relation
to the root of the tooth and its alveolus. My principal
object in presenting this illustration has been to give a
correct outline picture, including the entire root with its
alveolar walls, in which the direction of the fibers should
be correctly indicated in all of its parts. For this purpose I have selected a section in which the fewest number
of blood-vessels appeared, and in which there is, therefore, the least distortion of the fibrous arrangement. On
the lingual side, there is some modification of the form
of the alveolus occasioned by the nearness of the crypt
of the permanent tooth, a portion of the sacculus of
which is seen at *m*. This has caused the thickness of
the membrane to be diminished in its neighborhood.
Farther toward the crown, and also toward the apex of
the root, the membrane is thicker. A close study of the
illustration, with the aid of the description accompanying
it, will give a good idea of the general form and arrangement of the membrane.

THE PRINCIPAL FIBERS OF THE MEMBRANE.

Those fibers which are fixed in the cementum and
from thence stretch across, and are fixed in the alveolar
wall, or into some other tissue, as the fibrous mass of the
gums, and thus serve to maintain the tooth in its position,
I shall term the principal fibers of the peridental membrane. These are of first importance in the study of this
membrane, for with the exception of some deviations from
the usual course of these for the accommodation of the

blood-vessels and nerves, the other elements are so disposed as not to interfere materially with their arrangement. The structure is at the same time so very complex that we need to bring to our aid every available device for gaining a clear comprehension of the arrangement of its elements. To this end the arrangement of the principal fibers should be first studied, and afterwards the character of the fibers themselves, and following this the inter-fibrous elements.

ARRANGEMENT OF THE FIBERS.

Beginning with the gingival portion, we find the principal fibers firmly fixed to the cementum, literally springing out of it, and passing directly out, or but slightly divergent, from all the surfaces of this part of the tooth. The manner of the fixation of these fibers in the cementum will be studied in detail later.

On passing out from the cementum they may retain the solid form (fig. 39) or split up into fasciculi of finer fibers (fig. 42). In the latter case, which is the more common, they show some disposition, in many localities, to gather into loose bundles, the elements of which pursue a common course. But more generally perhaps the bulk of the fibers lie parallel with each other, deviating only to give place to blood-vessels and nerves, or the larger groups of lymphatics. Upon the labial and lingual surfaces of incisors these, after passing out some little distance from the tooth, are lost in the coarse, tangled, fibrous tissue of the gums. This is fairly well seen in figs. 36 and 40, g. g. Usually there is a fairly strong fibrous bundle turning down into the gingivus (fig. 36, h), especially on the labial side. Nearer the border of the alveolar wall the fibers pass on under the gum tissue proper, and are continuous with the outer layer of the periosteum of the outer surface of the alveolar walls. As these pass the margin of the alveolus, fibers, springing out of the bone, first decussate with, and then become mingled with them, thus forming a very firm support to the gingivus. This bundle has been termed the dental ligament.

As we pass around the teeth toward the lateral surfaces a disposition of the fibers to bend away laterally is noticed (fig. 40), and before we have fully reached the lateral surfaces the fibers may be traced continuously to the neighboring tooth, following a somewhat curved course, and passing the lower margin of the alveolar wall. Between neighboring teeth the fibers pass directly, or at a slight inclination from one tooth to the other, being fixed into the cementum of each (fig. 40, f). In the central part of their course many blood-vessels are seen, which cause more or less deflection in the course of individual bundles of fibers. If the horizontal section, including two teeth, is cut very close to the margin of the alveolar walls the fibers will be found to break up into bundles near the central portion, and many of them pass out of the section, while a portion continue on from tooth to tooth. The arrangement of the fibers on the lingual side does not differ materially from that of the labial as shown in fig. 40.

The gingivus, or free border of the gum (fig. 36, h) is covered with a moderately thick but very dense epithelial coating, surmounted upon the fibers emanating from the cementum of the neck of the tooth and the dense tangled mass of fibrous tissue forming the gums. Considerable importance has been given to that portion of the epithelium of the gingivus lying next to the neck of the tooth, which is composed of softer and more delicate cells than other portions. I will return to this point later.

The fibers of the lower portion of the body of the membrane run nearly directly across from the cementum to the walls of the alveolus, which they enter. Those entering at the rim of the alveolus usually have an inclination upward (toward the root of the tooth) as shown in the illustration, but a little farther upward they ruu squarely across, following nealy a straight course. It is here that the largest and strongest fibers of the peridental membrane are found, and we can often trace individual fibers entirely across from the cementum to the bone, even in young subjects, both in lengthwise and cross sections. It is the region from which the sections for figs. 37 and 38 were taken. As we pass farther toward the apex of the root the trend of the fibers in passing from the cementum to the alveolar wall becomes more and more downward (toward the crown), and at the same time a greater disposition to form into fasciculi or loose bundles is noted. In young subjects, when the membrane is thick, some of these bundles are very long, reaching for a considerable distance along the root to be attached finally to the alveolar wall. This disposition to form into fasciculi is seen most prominently perhaps, well up toward the apex of the root, where a greater portion of the alveolus is occupied by different tissues. Here fan-like fasciculi radiating from the cementum bone quite common, standing out as broad bands of fibers pursuing a straight course from the cementum to the bone, as if put upon the stretch ; as shown on the lingual side near the end of the root of the tooth in fig. 36.

Finally in the apical space (fig. 36, *e*) the disposition of the fibers is extremely irregular. Indeed in young subjects the tissue here has often more the appearance of embryonic tissue, in which there are few fibers developed, except those accompanying the blood-vessels, which latter are large and often divide into a number of branches, some entering the apical foram to supply the pulp of the tooth, and others passing down in the peridental membrane. In older subjects fibers are developed here which pass pretty directly from the root to the bone in radiating fasciculi.

This is in brief the arrangement of the principal fibers of the membrane, as seen in well prepared sections. However, many sections, when large numbers are examined, will come under the lens, which show wide variations from this arrangement. First, if the lengthwise section is cut a little to one side of the center, perhaps very few fibers will appear; or one not well skilled in the examination will fail to make them out as they lie among the cellular elements, for the reason that they are cut across at more or less of an angle. Especially will this be the case if the section is mounted plain in glycerine, or if it be well stained with a good nucleus-tinting dye. In either case the cellular elements will be rendered prominent, and the fibers, which remain transparent, will be hidden from view. I have many sections in which no fibers could be made out, but from the fact that they are so thick that the cells of the inter-fibrous tissue, which lie between them, appear in rows (fig. 41). Those prepared in the same manner, but which are a little thicker, so that the cells lie upon the fibers as well, show no appearance of fibers whatever. The same class of sections, however, stained diffusely with carmine, or pigmented, show the fibers prominently. Secondly, many lengthwise sections follow blood-vessels which traverse the membrane in this direction. These often are surrounded by more or less indifferent fibrous tissue. These fibers may lie parallel to the course of the vessels, and it is manifest that where the line of one of these happens to be followed, the membrane will appear, or rather will actually be divided into two parts, one of which will be attached to the cementum, and the other to the alveolar wall. It will be seen at once that unless the elements seen are well understood, a very erroneous conclusion may be arrived at, that of a double membrane. One would suppose from reading our literature that such had actually been the case.

With increasing age the cementum is thickened, the

walls of the alveolus are strengthened, the thickness of the peridental membrane diminished, and all its fibers shortened by being included in the cementum on the one side, and in the bone on the other. In this case the fibers generally appear to pass more directly from the cementum to the bone in all of the upper portion of the alveolus; yet the general trend, as illustrated in fig. 36, is fairly maintained.

If this arrangement of the fibers be studied with reference to the physical functions of the membrane—*i. e.*, that of maintaining the tooth in its position during the strain of its normal usage, it will be found that it is the very best that could be devised for the purpose. The tooth is swung in its socket in such a manner as best to resist a strain upward upon its crown, and save the tissues of the apical space from injury, while the fibers running squarely across in the lower third of the body of the membrane prevent displacement laterally.

The fibers of the peridental membrane are wholly of the white or inelastic connective tissue variety. There are no elastic fibers, or, at least, I have not been able to find any remaining after dissolving out the white fibers with alkaline solutions. The *form* of the principal fibers, though in many respects bearing a close conformity to the fibers of the internal layer of the attached periosteum is peculiar to this membrane. Indeed, in many localities no difference could be observed, if the examination were confined to the immediate surface of the alveolar process. The fibers of this portion are, however, in the main larger than those of the periosteum and rather less thickly placed. While in many localities they break up into fine fibers almost immediately after passing out of the bone, as is the case in the inner layer of the attached periosteum, in others they continue far out into the membrance as strong, seemingly, solid cords, with perhaps finer fibers and cellular elements running in a different direction, interwoven between them (fig. 46). In thin sections cut parallel with the fibers and stained with a good nucleus-tinting dye this arrangement gives the tissue a very characteristic appearance. The fibers are, in this case, perfectly transparent and invisible, and the cellular elements, which lie between them, appear disposed in rows as shown in fig. 41. Sections from the same series diffusely stained, or pigmented, show the fibers prominently.

The appearance of the fibers varies very much in different cases, and even in the same case in different modes of preparation and staining. For instance, fibers that appear as solid cords when stained with hematoxylin, may appear as fasciculi composed of very fine fibers, when stained diffusively with carmine. By this means we learn that the most compact of these coarser fibers are really condensed bundles of fine connective tissue fibers. In the peridental membrane these, the individual fibers, are often as compact as any that I have seen from the strongest tendons, but these coarser fibers are not themselves gathered into bundles as in the tendons. Hence the fibrous arrangement here is not similar to that of a tendon or ligament. While the fibers in many localities, especially next to the bone, are large and strong, each one stands somewhat apart from its neighbors with other elements intervening, which is not the case in the tendon or ligament. Therefore while the passive function of this membrane is that of fixation of the tooth in its position and of the same nature as that of a ligament, it has not in any of its parts the structure of a ligament.

The fibers passing out from the cementum are somewhat smaller and more thickly placed than those springing from the alveolar wall, but otherwise have the same character. In young subjects the parallel arrangement of these is somewhat interrupted near the cementum by the lymphatics which lie between them in great abundance. These fibers springing from the cementum on the one

hand and from the bone forming the walls of the alveolus
on the other, stretch across the intervening space in the
directions indicated in fig. 36, but there is something more
in the arrangement. While in some cases individual
fibers are maintained as such, and can be traced from side
to side, as shown in fig. 39, the rule is that the larger
fibers springing from either source break up into fasciculi
of very fine fibers as shown in fig. 42, and their individuality is lost by commingling with others. The fasciculi
are, however, continuous from the bone to the cementum.
The central part of the membrane therefore seems, and is
actually composed of finer fibers than either its cemental
or osteal margin. The fibers springing from the cementum very generally break up into fasciculi almost immediately. It is only near the cervical border of the
membrane or opposite the rim of the alveolus that we see
them generally continuing as solid fibers for any considerable distance. However occasional fibers, or a few
together may be found here and there in any portion of it
which do continue from side to side, especially near the
apex of the root where there are often found individualized groups of such. Those arising from the bone and
especially those near the rim of the alveolus often continue as solid fibers through one-third or even one-half
the thickness of the membrane, giving off only occasional
small fibers, but near the central part they generally break
up into fasciculi and their identity is lost.

In the thick membranes of young subjects there is
usually a very distinct vascular region near the central
portion; or midway between the bone and the cementum.
It is in this region or zone that the principal blood-vessels
and nerve-trunks are found. This causes much irregularity in the course of the fibers of this region, for the
fasciculi are deflected from their course in passing the
vessels. Besides these deflections we often find considerable bundles of fibers, especially in the middle portion,

or high up on the root, pursuing a course very different
from the general trend, and these pass between the fasciculi and have the effect of modifying their course. In
cross sections of the contents of the alveolus these are
seen cut across.

As age advances the appearance of the tissues of the
membrane is considerably changed. Most of the cellular
elements disappear, and the fibers appear more prominently. These latter are very much shorter, and it is
easier to follow them through from the cementum to the
bone. In many regions there is no appearance of interfibrous tissue whatever, and fasciculi or bands of fibers
cut parallel to their length often appear prominently. In
fig. 39, I have represented such a band of fibers seen passing from the cementum a, to the alveolar wall b. The
illustration is taken from a perpendicular section of the
roots and alveolus of a first molar of a man seventy years
old. The fibers of the lower portion of the figure c, pass
entirely through the membrane without breaking up into
fasciculi. This occurs only occasionally in rather small
bands of fibers lying parallel or nearly so. The more
general form of the fibers is that represented at d, in the
same figure where they break up into fine fibers soon after
leaving the cementum, as in figs. 42 and 43. In passing
the sections about under the lens, bringing different portions of the membrane successively into view, a great variety of appearances will be noted. Wherever the section
is parallel with the length of the fibers they will be seen
emerging from the bone and cementum, and generally
breaking up, as shown in fig. 42, into fasciculi that then
pursue a wavy course, usually more or less obliquely
toward the other side. Not very infrequently strong
groups will be found arising from either bone or cementum that spread out fan-like, as seen in fig. 43, some of
which may be traced through the membrane while others
pass out of the section or are lost by mingling with other

fibers. In rambling over the membrane with the microscope many points are found at which there appears to be no attachment whatever. This in many instances, is from the fact that the section is not parallel with the fibers at that point, which can often be definitely made out by the presence of fibers cut obliquely. But at other points, and these are not few, certainly no attachment has existed. Indeed the principal fibers may be absent and the fibers of the indifferent tissue may lie flat upon the surface of the cementum or bone for a considerable space. Their course being parallel with the surface. At some points the cause of the detachment is evidently absorptions of the alveolar wall or of the surface of the cementum, and the principal fibers may be seen to have been severed. The osteoclasts are found frequently, and the roughened surface tells plainly of their action, even at this advanced age. It has become plain to me after a long study of this point that the attachment of the fibers is continually changing. They seem to be loosened and remain so for a time, but they are again attached, or new fibers are developed. Some other part is loosened and again attached, so that in passing around the root of a tooth of an old person these non-attached points are continually coming into view. This will be studied more in detail under absorptions within the alveolus.

As has been said, the interfibrous tissue is much diminished in old age. At many points none whatever can be seen ; at some points, however, there is so much of this that it might be mistaken for a young membrane. In running along the surface of the cementum occasional groups of cementoblasts appear with undeveloped cells in their neighborhood. The same appearances of local activity are met with along the bony wall also.

CHAPTER IX.

Other than the blood-vessels and nerves, the principal interfibrous elements are an indifferent tissue, and the various forms of cells. The principal fibers are accompanied by fibroblasts, which belong to them, and are rendered prominent by any nucleus-tinting dye. Fig. 50, D. But aside from these there is among the principal fibers a very considerable number of fibroblasts, accompanied by very fine fibers, which pass between the principal fibers and often pursue an independent direction. Fig. 46. This I have termed the interfibrous or indifferent tissue. It seems to pervade the entire membrane, and is found wherever the principal fibers are absent, or are coarse enough for it to be distinguished. In figure 46 I have represented this tissue with a high power. The illustration is taken from the margin of bone near the rim of the alveolus, where the principal fibers are very large. The interfibrous tissue is seen to be ordinary fibrous connective tissue containing the usual fibroblasts. In this instance its course is diagonal to the principal fibers. This tissue is well seen in many regions of the membrane, but when it is mingled with the principal fibers of the vascular region its identity is necessarily lost, except where it forms an investment for the vessels and nerves. Fig. 50, E and F. Parts of the membrane here and there are made up seemingly of this tissue, the principal fibers being absent. Fig. 49 d. This is most frequently seen high up on the root or about its apex, and marks especially those regions of the membrane that I have designated in previous pages as non-attached. In general, when standing alone, it gives the appearance of indifferent tissue. In the young subject its fibroblasts are a marked feature of the membrane when stained with a good nucleus-tinting dye, but in the old these become very thin scales, and do not take stains well, so that it is difficult to distinguish them except with high powers. The tissue then appears loosely fibrous, and the fibers, while pursuing no very definite direction, have a general tendency to lie horizontally to the cementum, or parallel with its margin, as seen in lengthwise sections. It does not seem to attach itself to the cementum or bone, as do the principal fibers. The fibrous investment of the blood-vessels and nerves, additional to the tissue properly belonging to their walls, seems to belong to this tissue. In young subjects this is often very abundant (fig. 50, E and F), forming masses accompanying the vessels, and causing deviations in the course of the principal fibers. In older subjects this accompaniment of the blood-vessels mostly disappears, and in sections their walls look comparatively thin.

BLOOD SUPPLY OF THE PERIDENTAL MEMBRANE.

The blood supply of the peridental membrane is very bountiful in the young subject, and though it is much diminished with the thinning of the membrane as age advances the vessels remain fairly abundant. In the young subject there is a very well marked vascular area lying centrally between the cementum and alveolar wall, or often rather closer to the cementum. This is most regularly seen in cross sections. Through this portion the fibers are deflected from their regular course in many parts to give space to the larger arteries, veins and nerve bundles. Fig. 50, E. and F. The larger arteries enter the alveolus mostly at the apical space, or rather one or two vessels enter here, and immediately break up into

smaller ones.   One or two of these enter the root canal
to supply the pulp of the tooth, while the others—from
four to six or eight—pass down along the sides of the
root to supply the peridental membrane. In their passage
down the membrane these divide into many branches, a
considerable number of which enter the Haversian canals
of the alveolar wall, or receive branches from that source.
This kind of connection between the circulation of the
tissues outside of the alveolar wall and the peridental
membrane is very rich in young subjects, and although
the bone becomes much more dense with advancing age
it is still fairly well maintained. Seemingly for this reason
there is not much diminution in the size of the main
arteries passing down from the apical space to the gingivus, though they are much increased in numbers.   We
may therefore say very justly that the blood supply of
this membrane is received largely through the walls of
the alveolus. Some vessels are continuous, however, from
the apical space to the gingivus.   I have a few sections
from the incisor teeth of the dog cut lengthwise, inject-
ed, that show arteries traversing the membrane from
apex to gingivus without break, but giving off and receiv-
ing branches from the alveolar wall throughout their
course.   Many of these branches can be traced through
the alveolar wall to their connection with the larger vessels of the gum-tissue.   These vessels during their course
in the peridental membrane give origin to a fairly rich
capillary plexus that supplies its tissues.   It is rare to see
a vessel of any size close to the cementum.   This portion
of the tissue seems to be supplied almost solely, but very
richly, by the smaller capillaries.   The passing and re-
passing of the vessels through the alveolar wall is well
seen in sections, and shows many of the larger ones near
the bone and within its canals. With the thinning of the
membrane as age advances the vessels are found lying
very close to the bone, and in case the membrane is very

thin many of them lie in grooves in the bone. This is best seen in cross sections. (Fig. 37.) Veins accompany, or perhaps stand a little apart from, most of the larger arteries. At the rim of the alveolus the vessels of the peridental membrane anastomose very freely with those of the gum, and this gives a pretty rich gingival plexus.

From this arrangement of the circulation of the blood in this membrane it will be seen that it is not readily robbed of its blood supply by accident. In case of alveolar abscess involving the apical space, the blood supply from that source is cut off, but that through the alveolar wall and by way of the rim of the alveolus is ample; or even in case the supply from both the apical space and by way of the rim of the alveolus is cut off simultaneously, the remaining body of the membrane will be supplied through the alveolar wall, and will not suffer from want of blood. An inflammation involving one part does not necessarily endanger another.

The sensory function of the peridental membrane is supplied by nerves entering it in company with the bloodvessels, and from all the sources of blood supply mentioned under that head. The principal bundles, however, enter by way of the apical space, and then divide, a portion entering the apical foramen for the supply of the pulp, while the others pass down the sides of the root supplying the peridental membrane. A considerable number enter through the walls of the alveolus by way of the Haversian canals, each containing from four to ten or more nerve fibers. These traverse the membrane, giving off smaller bundles which are lost in the tissue, until the gingival border is reached, where, in company with those of the gum tissue, a rather rich plexus is formed.

Specialized nerve terminations have not been found in this membrane in sufficient numbers to show that they are essential. I have seen a few Pacinian corpuscles near the gingival border, and rarely some other knob-like terminations. Generally, however, none of these are found. The bundles of fibers sub-divide into single filaments which are lost in the tissues, and probably terminate mainly as naked fibers.

Through this supply of nerves the peridental membrane becomes the organ of touch for the tooth. The enamel, the portion of the tooth exposed, has not the sense of touch. This may be demonstrated by experiment in many ways. One of the simplest of these is performed as follows: Take any small instrument and touch .with it the enamel of any tooth. It will be found that the lightest touch is felt distinctly by the patient. Now place the finger on the opposite side of the tooth and make firm pressure, and while this is maintained again touch the exposed part of the enamel with the instrument. It will be found that under these conditions the touch will not be felt. Now by varying the pressure with the finger it will be found that in order for the touch of the instrument upon the enamel to be felt it must be sufficient to overcome the pressure brought to bear by the finger. A slight movement of the tooth must be produced so that it may effect the peridental membrane. without this no sense of touch is manifested. This is different from the temporary semi-paralysis that may be produced by firm continued pressure, or by a blow. In order for this to be effective it must be pretty severe or long continued. For comparison this may be tried upon the hand or fingers. This simple experimentation readily demonstrates that the sense of touch in the tooth is very different from that of the skin. The sense of touch in the finger nails will be found similar to that of the tooth.

Normally, the sense of pain is not easily aroused in the peridental membrane. The office of fixation of the tooth, and maintaining this against the heavy pressure normally brought to bear upon it, demands the capability of withstanding heavy strain and blows without complaint, and at the same time without limiting, in the absence of such a strain, the acuteness of the sense of touch. The membrane does not, however, on this account, bear mutilation without pain, but is, perhaps, as painful as the average of the tissues, and in its inflamed state it becomes exceedingly sensitive to very slight pressure, as is uniformly witnessed in acute pericementitis. Its rich supply of blood-vessels and nerves renders it capable of rapid recovery from injuries of almost any kind. Indeed, there is no tissue of the body that shows a more marked tendency to recover from severe injuries.

These sensory functions are not destroyed by injury to any particular portion of the membrane. I have carefully tried the sense of touch in teeth after having removed all of the contents of the apical space, i. e., after soreness had so far abated that the sense of touch was not abolished by the sense of pain, and found the sense of touch was, as far as could be ascertained, normal. That its sensitiveness to painful impressions is not abated by the destruction of the nerves entering by way of the apical space, is sufficiently obvious to all who have had to do with large acute alveolar abscesses producing extensive destruction of the tissues of the apical space. It follows, therefore, that the nerves entering the membrane through the walls of the alveolus are sufficient for the maintenance of the sensory functions.

CHAPTER X.

The peridental membrane has a very peculiar system of cells closely resembling those of the lymphatics. In young subjects these are found in great profusion lying among its fibers close to the cementum. I know of no other system of cells similar to this anywhere within the bodies of men or of the lower animals. They seem as distinctly specialized as the agminated glands or Peyer's patches of the small intestine. I therefore regard them as peculiar to this particular portion of the peridental membrane. They occur mostly in the form of rows of cells insinuated between the fibers of the membrane. They are never far from the cementum, but not in contact with it, except in some isolated cases observed in the pig. These rows of cells anastomose freely with each other and form a network over the whole of the root of the tooth. Their number is so great that I have counted from one hundred to two hundred of them cut across in the cross section of the root and alveolus of an incisor tooth of ordinary size. Fig. 50, C, C. These rows of cells vary very much in the numbers of cells included in their make-up. Sometimes a cross section will show only one or two cells lying together. Again, and more commonly, five or six that form a rounded group, and more rarely, especially near the gingivus, where they are generally larger and more numerous, there will be quite a body of them giving with high powers a gland-like appearance. I have represented one of these in fig. 47, using for the purpose the one-twelfth inch objective, in which its relations to a small capillary vessel are shown.

Very often the cells lie between the fibers in such a way as to show, in cross or lengthwise sections, rows running outward from the cementum. This is especially well seen in the pig. These often seem to be single rows of cells, or they may consist of two or three rows lying side by side. In either cross or lengthwise sections the islands of cells seem to be entirely detached from each other, especially if the sections are very thin. But in sections cut horizontal to the surface of the cementum at such a distance as to include them, they are seen to be in the form of chains that anastomose with each other, like a network. In fig. 48 I have represented a group of these, using the one-eighth inch objective. This is very readily seen with low powers if the section, cut in the manner indicated above, be double-stained with carmine and hematoxylin. In this case the lymph cells take the hematoxylin and the fibrous tissues are stained red, and with low powers, the first impression will be that of a fine capillary injection. Higher powers will reveal the true character of the tissue.

The individual cells are like those of the lymphatic glands. They show a circular or polygonal outline, and the central portion takes the staining agent strongly. In the larger groups it is easily seen that they are enveloped in a very delicate limiting membrane. This limiting membrane is not so easily seen about the smaller groups or rows of cells. However, by following one of these carefully, which I regard as a very delicate lymph duct crowded with lymphoid cells, I have been able in many instances, to connect it with the smaller veins or capillaries in the form of the perivascular spaces peculiar to the lymphatics of other regions. Owing to the extreme difficulty of obtaining nitrate of silver stainings of this membrane, it is specially difficult to make out these points quite satisfactorily.

Klein seems to have shown that nodes of lymph cells develop within the lymph sacks or enlargements of the

lymph ducts, which he designates as endolymphangeal,
and also outside, but in contact with these, which he
terms perilymphangeal nodes. These are, however, endolymphangeal, as distinguished from the perilymphangeal
glands or nodes. In fact, these seem to be lymph canals
that are packed with lymphoid cells rather than true lymphatic glands. The cells are very well seen in plain
glycerine mountings, especially after acetic acid, and the
groups may readily be made out with the half-inch objective, but as they lie crowded among the other tissues
higher powers are necessary to differentiate them.

These cells are more abundant in the omnivora than
in other animals that I have examined, and the pig is an
especially good subject for their study. They are very
well seen in the herbivora, also, but seem not so abundant
in the carnivora. They seem to diminish in numbers as
age advances, though this point has not been studied sufficiently. In one membrane from a man, forty years old,
the number seems to be much diminished, though groups
of them were seen in almost every field. In another from
a man about seventy, only a few groups of the cells were
found. It seems probable that they disappear, for
the most part, with advancing age. In this they agree
with specialized lymphatics elsewhere, such as Peyer's
patches of the small intestine, and a few that have been
noted in other positions.

One circumstance, aside from the histological interest,
has directed my attention quite strongly to these cells. In
extracting a cuspid tooth a large piece of the anterior portion of the alveolar wall broke away, adhering to the root
of the tooth, and gave me the opportunity of making sections for the study of its membrane. Phagedenic pericementitis was destroying the membranes of some of the
other teeth, but about this one no pockets were observ-
able, though there was some slight redness of the gingivus.
On microscopic examination, I found that some of the

lymphatics near the gingival border of the membrane were in a state of suppuration, while some others did not take staining agents well. This condition followed the lymph chains in the direction of the apex of the root to a distance that surprised me, considering the very slight signs of disease visible before operating, and seemed especially confined to these cells. Examination of them for micro-organisms was not thought of in time.

This case hints quite strongly that these lymphatics are the seat of this very peculiar affection. It seems that it is also these glands that are first affected in salivation with mercury, when, as physicians say, "the gums are just touched," and the teeth become sore, when pressed together. Also, when the teeth become sore from other causes that may be regarded as constitutional, i. e., from some agent in the blood that affects these glands.

Formerly, it was suggested by Serres, who is quoted by Salter, that the inner portion of the epithelium—that portion clothing the border of the gingivus folded in against the tooth—acted the part of a gland. This part of the epithelium is softer than other portions, and the gland-like action noticed here under the influence of iodide of potassium, mercury, and some other remedies, evidently led to this conclusion, which, from clinical observation, seemed to me to be justified. (See American System of Dentistry, vol. I, p. 955.) But since I have made a more critical study of these lymphatics, it has become clear that the results were derived from them which had been attributed to the inlying epithelium of the gingivus. I find the lymphatics to be larger and much more numerous just in that neighborhood, and while I have found no such thing as a duct leading to the gingival aperture, the glands lie in very close proximity to it. Furthermore, a portion of the connective tissue in immediate conjunction with the tooth is not covered by the epithelium. In other words, there is no attachment of the epithelium to the

root of the tooth. It seems to be through this space that
the cells—so-called salivary corpuscles—found under the
free border of the gingivus, pass. These may be found
at any time under the healthy gingivus, and their numbers
are augmented with every irritation of the membrane.
Indeed, close clinical examination makes it apparent that
there is a slight secretion at this point that is not quite
satisfactorily explained even yet by microscopic study of
the part.

HARD FORMATIONS WITHIN THE PERIDENTAL MEMBRANE.

There are occasionally found, in the tissues of the peridental membrane, especially in elderly persons, certain
hard formations that resemble the calcospherites so frequently found in the tissues of the dental pulp. I have
seen more of these about the roots of the molars than
elsewhere, but have also found them along the sides of
.the roots of the bicuspids. Occasionally I have seen
these built into the substance of the cementum, especially
in hypertrophies.

In fig. 49 I present an illustration of one of these,
rather a small one, from the membrane of a bicuspid,
which presents their usual appearance very fairly. They
are composed of concentric rings of lime salts united by a
basis substance which appears identical with the phlebo-
liths of varicose veins, and calcospherites of the dental
pulp. They do not, however, present exactly the same
features of either of these. They are much larger than
the calcospherites of the dental pulp, and the incremental
bands or layers are much thicker. Neither do they present the nodular forms composed of numbers bound
together in one mass, so common to either of the before-
mentioned bodies. They are usually seen like the one
in the figure, as isolated spherules, many of which are
large enough to be readily seen with the naked eye. In
several instances I have seen two spherules united, but

this is unusual. The larger ones, as they appear in sections, after being decalcified, of course, usually show cracks radiating from the center toward the circumference. This may have occurred in the cutting, though I have no means of determining it. I only know that the smaller ones usually show nothing of this kind. In several instances I have seen cementum built upon the larger ones, and fibers attached, showing that they may become a nidus for an irregular or nodulated hypertrophy. I have also in my collection hypertrophies which show these forms in their substance. Further than this, I know nothing of the origin or significance of these bodies.

In the peridental membranes of old people there is a considerable number of pigment granules found. They occur isolated, or in groups, oftenest about the walls of the blood-vessels, but also a part from these. They are intensely black, rather small, and seem to be amorphous. They remind one very much of the pigment so generally seen in lung tissue. They are not present in the peridental membrane of young persons. Nothing is certainly known of their origin or significance. The idea that they arise from extravasations of blood might be suggested, but this seems improbable.

CHAPTER XI.

The osteoblasts of the peridental membrane are found on the inner surface of the walls of the alveolus in abundance in young subjects. They lie on the bone between the principal fibers (figs. 45 and 46), and there are generally many young cells in the neighborhood, filling in between the fibers if they are large and solid as in fig. 46, or in the meshes of the finer fibers when they break up close to the bone as in fig. 45. But in this respect the utmost variety will be found. Many localities, even in young subjects, will be found almost destitute of these cells, while others, at only a little distance perhaps, will be crowded with them. In aged subjects they are generally absent, or are represented only by very thin flattened scales, lying close against the bone, that are very difficult of observation. But even in these cases occasional areas will be found in which the osteoblasts appear, covering the bone as in the young, only less profusely. These are undoubtedly areas of activity, points at which bone is being built up to accommodate some change in the position of the tooth. This will be discussed farther under the head of absorptions taking place in the alveolus.

The building of bone occurs on the inner walls of the alveolar processes in that growth which fits them about the roots of the teeth. These additions are made in the same manner as subperiosteal bone is built up under the attached periosteum, to which the reader is referred. The peridental membrane is very thick in young subjects, and the alveolus correspondingly wide. Bone is deposited upon the inner walls of the alveolar process as the membrane is reduced in thickness. Indeed after the alveolar process is once formed the subsequent deposit of bony matter is mostly on the inner side, filling in the enlarged space through which the crown passed in the process of eruption, to conform it to the root of the tooth. In this growth of bone new canals seem not to be formed by the growth of processes which arch over, as in subperiosteal growths. Nearly all the canals formed open into the alveolus very nearly in the direction pursued by principal fibers of the peridental membrane. (See Fig. 36.) Many of these approach the alveolus so obliquely that in cross sections such canals appear. In lengthwise sections, however, their true character is sufficiently apparent. The bone is therefore built up in the first instance in the same manner as solid subperiosteal bone, but with canals running in the direction of the growth, or at an angle inclined to that direction. This growth of bone shows the residual fibers very plainly in many of its parts, for the fibers of the peridental membrane are included in this in the same manner as the fibers of the attached periosteum. This I have attempted to illustrate in Fig. 51, from a perpendicular section through the rim of the alveolar wall, choosing the extreme point of the alveolar process — that represented by b, Fig. 53 on the labial side; d, Fig. 51 represents the extreme point of the rim of the alveolar wall. f, f, The subperidental bone which is closely filled with residual fibers from the large fibers of the peridental membrane, b. Subperiosteal bone, which is usually small in amount, and on the labial side, is confined mostly to the immediate rim of the alveolar wall; for on the labial side there is more often found absorption of bone, thinning the alveolar wall as it is built up on the inner side. The Haversian bone is left without stippling that it may be more plainly marked, and is pointed out by a. g, g, g, Are points at which absorption of bone is in active progress. e, Points out the fibers of the peridental membrane. This bone, forming the alveolar wall, it will be seen, is first built up solid as under the firmly attached periosteum. There is, therefore, no difference in the building of bone here and elsewhere, except that the included fibers are larger, which gives the bone quite a characteristic appearance. This bone is very soon invaded by absorbents, and canals are burrowed through it, which is followed by the deposit of systems of Haversian bone, thus removing the fibers, as shown in the figure. This process follows very closely the building of the bone, so that there is not at any time a very considerable amount of the alveolar wall that shows residual fibers. This is well shown in Fig. 51, in which I have left the Haversian bone without stippling to distinguish it more clearly. At maturity the bone has become so changed by this process that the residual fibers are confined to the immediate surface, and almost the entire mass of the alveolar wall is seen to be made up of secondary Haversian systems. In old subjects these show all along the inner border the effects of absorptions and rebuildings of bone that have occurred from time to time for the accommodation of changes in the positions of the teeth. This matter will be discussed in detail later.

A description of the origin of the alveolar process belongs rather to embryology, and I shall not enter the discussion of that part of the subject here. The growth of the alveolar process, after the tooth has taken its place in the arch, presents some peculiar features. This growth is, in a large degree, contemporaneous with the development of the tooth's root, whether it be a temporary or permanent tooth. The socket at this time is usually much too large for the root, and the peridental membrane is correspondingly thick. This is necessarily the case in the first instance, for the accommodation of the fully formed crown of the tooth. After the tooth has taken its position the alveolus grows smaller by the deposit of bone on its inner wall, until it is brought more nearly to the size required by the root which it is to support. This occurs very rapidly as the tooth is taking its position in the arch. There is, however, a movement of the permanent tooth, after it has taken its position in the arch, to which I wish to call special attention. This takes place largely during the very noticeable change which occurs in the features about the age of puberty, but is in progress from the time the permanent incisors have taken their places until maturity. I have illustrated this movement diagramatically in figs. 52 and 53. In each figure an incisor tooth is represented in dotted lines with the rim of the alveolus at a, a. e, Represents the apex of the root, and the dotted line c the inner wall of its alveolus; while the space between the lines c and d shows its thick peridental membrane. This represents the position of the tooth and its alveolus at the age of ten or twelve years. The tooth and alveolus drawn over this with solid lines represents the same tooth in the position it will have assumed at the age of twenty-one or two years. The growth of the alveolar process has carried the tooth in the direction of its length about the distance represented by the length of its crown, or that part of the tooth covered by enamel, as represented in fig. 53, which is the maximum movement that I have observed, while the minimum movement is about one-half the length of the crown, as represented in fig. 52. This movement seems to be rather greater in men than in women, but it presents considerable variation. I should say that the various points of growth of the bones of the face are possibly not determined yet with sufficient accuracy for fixed points to be established that will be without objection. The measurements I have used have been from the anterior spinous process of the superior maxillary bones (figs. 52 and 53, J), to the cutting edges of the superior central incisors. This measurement, made at ten or twelve years of age and at maturity,

indicates the movement shown in the diagrams. This is
a large factor in the elongation of the face. The movement in the lower jaw seems to be about the same as that
in the upper, though it can not be so definitely determined. The principal growth concerned carrying the teeth
forward for the elongation of the arch, is, as is well
known, at the back part of the maxillæ, carrying the anterior portions forward to make room for the molars.

The growth of the alveolar process, which carries the
tooth with it, as shown in 52 and 53, is almost entirely
from the osteal side of the peridental membrane, or upon
the inner side of the alveolar wall. The elongation is
made, it is true, upon the rim of the alveolus, the portion
represented in fig, 51 growing in a line almost parallel to
the length of the tooth from the points a, a, in figs. 52
and 53, to the points b, b, and in the meantime all of the
space between the wall (inner) of the former alveolus c,
and that of the final alveolus f, is filled in by growth of
bone from the osteal side of the peridental membrane.
This is all built in originally with the character of bone
represented at f, fig. 51 and in figs. 45 and 46, and is removed by absorption and replaced by Haversian bone, as
represented at a, fig. 51. The plan of this removal and
rebuilding is more particularly described in fig. 24, and
on page 000, to which the reader is referred. In this way
there is a continuous activity in growth and reconstruction of the alveolar processes during this time, in which
the tooth itself, except its cementum, is passive, the
dentine and enamel having previously completed their
growth. The movement represented in these figures is
seen to be almost wholly in the direction of the long diameter of the tooth, but there is some movement of the
crown of the tooth forward in the direction of its short
diameter. This is accompanied by a tilting of the crown
forward, as shown. I have often found absorption in progress about the point h, and observation seems to indicate

that a reduction occurs as represented from the line h to g. However, from the want of a fixed point from which to measure, it seems almost impossible to determine the amount of the movement in this direction. The tilting of the crown forward is readily determined, however, by taking the relative positions of the tooth to a perpendicular line.

I do not know that any previous writer has discussed this subject, and I have not now sufficient data at hand for the full presentation of it. Yet it is of great importance in connection with the formation of the dental arch, and serves to illustrate the necessity of retaining it complete during the formation of the features. It also has an important bearing on the subject of the correction of irregularities. I will have more to say of it after having considered the cementum.

CHAPTER XII.

THE CEMENTUM AND CEMENTOBLASTS.

The cementoblasts or cement builders are to the cementum what the osteoblasts are to the bone. They are the cells concerned in the formation of the matrix, and the deposit of lime salts, which enter into the formation of the cementum. These are cells of rather large size and of peculiar form, and are found lying between the principal fibers of the peridental membrane and upon the surface of the cementum. While functionally they hold the same relation to the cementum that the osteoblasts hold to the bone, they have no resemblance to the osteoblasts in form.

The osteoblasts are polygonal cells inclining to the round form, and their longest diameter is often directed away from the bone upon which they lie, as has been said upon another page. I have never seen the cementoblasts presenting these forms, but on the contrary, they are always distinctly flattened cells with one of their flat sides resting upon the cementum. They are, indeed, in the form of somewhat thickened scales, of very irregular outline. This irregularity of outline seems to be due to the position they occupy among the principal fibers of the peridental membrane as these latter pass out from the cementum.

There is usually a central mass, which is seen to contain a regularly formed nucleus, and from this central portion irregular projections extend among or between the fibers of the neighborhood.

The cells, with their projections, are so placed that they occupy all the surfaces of the cementum, except that occupied by the fibers that emerge from it. A better idea of their form can be gained by examination of the illustrations, figs. 54 and 55.

In the first of these I have isolated several cells, and it will be seen that the form of the projections from the cell body is such as will fit in between the fibers. In the other illustration the fibers are shown cut across and left white, so that their outline may be better seen. These illustrations are taken from sections cut horizontal to the surface of the cementum, and are double stained with hematoxylin and neutral carmine, which gives a diffusive red stain to the fibers, while the cells are of a deep blue.

Sections cut in any other direction will fail to give a correct impression of the form of these cells, but the sections cut perpendicular to the surface are valuable as illustrating fairly the thickness of the cells. In such sections the cementoblasts are seen in parts only. In a given focus of the lens, isolated parts of the same cell may appear as small cells separated by fibers, and it is practically impossible to connect them and gain definite information of their form from such sections alone. The projections among the fibers of the membrane spoken of above are not in any proper sense processes from these cells, but are to be regarded as portions of the cell body which takes this form on account of the presence of the fibers.

I have made out true processes proceeding from these cells in but few instances, but enough to show that they exist upon a considerable number, if not all, passing into the cementum upon which the cells lie. However, they are evidently not so numerous nor so regular as the processes of the osteoblasts, or if so they are much more difficult of observation. I have never seen processes extending from these cells in a direction from the cementum out into the tissue of the peridental membrane. I think it probable that such processes exist, but it is imprac ticable to display them by stretching the tissue away from the cementum, attached as it is by strong fibers, in any manner similar to that represented in figs. 17 and 18, in case of the non-attached periosteum.

In the growth of the cementum some of the cemento-blasts are included in its substance, and persist as cement corpuscles in the same manner as the osteoblasts are included in the bone as bone corpuscles. The number and relative positions of these are, however, extremely irregular in those animals that have a thin cementum. About the necks of the human teeth and the teeth of the carnivora, there are usually no cement corpuscles, but at points where the growth of cementum is thicker, they appear in considerable numbers; and toward the apex of the root, where the deposit of cementum is considerable, they may appear in profusion. That regularity of occurrence which is noted in bone corpuscles, is not seen in the cement corpuscles. On the contrary, they appear in groups or in patches, while perhaps considerable areas are destitute of them. In some of the herbivora, and notably in the pig, they appear with more regularity, figs. 57 and 58.

The cement corpuscles have processes corresponding to those of bone corpuscles, but presenting great irregularities. Some may show none whatever, others a few that may be very short or very long. While others again have a great profusion that radiate in every direction, branch and anastomose with each other and with those of neighboring cells, forming an intricate network. Many of the corpuscles show processes passing in one direction only and that is usually toward the surface of the cementum.

The *cementum* is deposited upon the dentine and covers the root portion of the tooth. There is never an attachment of the soft tissues with the dentine upon its outer portion. Under some conditions the soft tissues may, indeed, lie in apposition with the dentine upon its surface, but there is no physiological union of the two structures. The physiological connection of the dentine is with the dental pulp, and upon the pulpal side of the structure. When the soft tissues lie in contact with the opposite side, whether during development or afterward, the physiological process is either the deposit of cementum upon the dentine, or absorption of the dentine.

The deposit of cementum is in the form of lamellæ, layers, or strata, and covers the root over its entire surface. These lamellæ are thin, normally, toward the neck of the tooth, and thicker, progressively, as the apex of the root is approached, the difference usually being very considerable. In normal conditions the number of lamellæ is about the same on all parts of the root, which gives a much thicker cementum at the apex than at the cervical portion of the root. The first of these, or at least the first part of the first lamellæ, is usually hyaline or irregularly granular and ordinarily contains no cement corpuscles, or at least but few. The next lamella, especially high on the root portion, presents these corpuscles very generally, and they continue irregularly through the successive lamellæ, provided always that the individual lamellæ be of considerable thickness. Very thin lamellæ, whatever their position, are usually destitute of corpuscles, while the thicker ones contain them.

These lamellæ seem to represent periods of activity in the deposit of cementum, each lamella being the result of a single period of activity. If we extract a tooth soon after its eruption and examine its cementum, we shall usually find it very thin and containing but one or two lamellæ. A tooth from a person who has reached maturity will present a larger number and the cement will be thicker than in that of a child of twelve or fourteen years, but not nearly so thick as the cementum upon the roots of teeth from old people ; nor will it contain so many lamellæ of cementum. These layers are subject to the greatest irregularity, both in the thickness of the single ones and in their number. Neither do they present much regularity at a given age in different persons. In all these respects there is the utmost irregularity. The individual lamellæ of cementum are divided by lines that may be very distinct, or but imperfectly seen. The mode of preparation makes much difference in the distinctness of these lines. Sections cut from decalcified teeth and mounted plain (without staining), in glycerine, show them very fairly, but they are rendered more distinct by tinting slightly with a diffusive carmine stain. These lines I will call, as Salter has done with good reason, the *incremental lines* of the cementum. This is appropriate from the fact that each one marks the divisions between the lamellæ that are laid upon the root, the one upon the other. Each successive lamellæ is younger than the preceding one, as we pass from the surface of the dentine outward.

In subperiosteal growth of bone, incremental lines occur similar to those in cementum, but they are rarely permanent, for, as has been said, subperiosteal bone is changed by the burrowing out of the bone first formed, and the deposit of Haversian systems in its stead. Nothing of this kind occurs in the cementum. It has no Haversian systems. In all of my examinations of this structure, I have not in any instance seen anything that could be called a Haversian system as these are known in bone. I have seen many canals that seem to represent small bloodvessels included in its structure, especially near the apex of the roots or between roots that have become fused by deposits of cementum, but these have never had about them deposits resembling the Haversian systems of bone. In normal conditions the lamellæ of cementum, when once deposited, are permanent. They may indeed be removed, or burrowed into, as I shall describe later, by absorptions beginning at the surface and cutting through the successive layers, but they bear no resemblance to the burrowing for the formation of Haversian canals in bone. Such absorptions are always refilled by a true surface deposit of cementum, if filled at all.   See fig. 61, a, a, a.

A correct understanding of these facts is important to the study of hypertrophies and absorptions of the cementum, which I shall introduce later. Furthermore, the cementum must, I think, be regarded as *continuously growing*, in the sense of not ceasing at maturity. It is very evident that its growth does not cease with the maturity of the tooth, nor with the maturity of the person. We find pretty uniformly a thin cementum upon the teeth of the young, and a thick cementum upon the teeth of the old ; and when a great number are examined from persons of known ages, it will be found that there is a continuous increase in thickness and in the number of incremental lines.   But in such examination great variations from any given rule will be noted.   One set of sections cut from the lower molar of a man, about seventy years old, shows on the sides of the roots forty-two lamellæ easily distinguishable and counted with a half inch lens.   While over the apex of the root, which presents some hypertrophy, there are a few additional lamellæ.

The incremental lines are not always regular in their distribution over the tooth's root.   Sometimes a lamella is laid down that covers only a part of the root and two lines merge into one.   This seems to show that there has been a local activity of deposit over part of the surface, that has not extended to the entire root.   Some of the incremental lines seen toward the apex of the root where the cementum is thicker may disappear as the neck of the tooth, where the cementum is thinner, is approached. Again, regions will be found in which it is evident that certain lamellæ of the cementum have been removed by absorption.

FIBERS OF THE CEMENTUM.

As I have said the fibers of the peridental membrane spring out of the cementum. These fibers pass through all of its lamellæ to the first one laid on·the dentine and part way through that, no matter what the thickness may be. In most localities in the human cementum these fibers are not continuous, but are broken at some of the incremental lines. At some such points they have certainly been detached by absorption, but in most instances this cause of detachment can not be made out satisfactorily. On account of this frequent breaking it is not generally possible to follow individual fibers from the peridental membrane entirely through the cementum, even in sections cut parallel with them, though they may be seen in all its 'lamellæ. In the pig the fibers are much larger and less thickly placed than in man. This renders the tracing of individual fibers from the membrane into this substance comparatively easy (Fig. 57). In the pig, also, there is a great thickness of cementum, comparatively, formed in a few months, and this presents but a few incremental lines at which the fibers are broken. We can, therefore, follow individual fibers through its entire thickness in sections cut parallel with them. In man, the fibers are so much broken at the incremental lines that it is only now and then that we are able to find individual fibers traversing its whole thickness.

Much of the cementum of man, especially that about the necks of the teeth, when so stained as to show them clearly, seems almost as if made up of fibers. These are usually small, placed close together, and run pretty squarely outward, pursuing a straight course, (Fig. 59, b, c,) but farther up on the root, where the cementum is thicker, they are often found curved in various directions, and many times we shall notice an abrupt change of direction at an incremental line. Some spaces or patches will be noticed in which the fibers seem to be absent.

These fibers have been noticed by various writers, and not a few have spoken of them' as the fibers of Sharpey, while others, Salter and Abbott, seem to have mistaken them for canaliculi similar to those of dentine. This error can scarcely be avoided if the examination has been confined to the dried specimen, for it seems that many of the fibers are but imperfectly calcified, and in drying suffer shrinkage to such an extent as to give that appearance. I have a number of sections in my collection that show this.

These are the principal fibers of the peridental membrane included in the cementum in its growth, and furnish the means of making firm hold of the peridental membrane upon the root of the tooth. They are white connective tissue fibers, the ends of which are included in the matrix of the cementum sufficiently to make them apparent when the lime-salts are removed, but when both are calcified, they can not be demonstrated except in cases in which there is imperfect calcification of the fibers, as has been mentioned above.

A very beautiful demonstration of these fibers may be had in the cross-section of them, i. e., in moist sections of the cementum cut horizontal to its surface. If these be very thin, stained, and mounted in balsam, they will show the fibers cut across especially well. In this case there will generally be such a shrinkage that a part of the circumference of the fiber will be parted from its matrix, showing it plainly; and by close focusing the whole outline of the fiber may be clearly seen. In some very thin parts of sections the fibers may drop out of their alveoli, leaving openings. This was the case in the section from which fig. 56 was made. In many very thin sections parallel with the fibers, we may see about broken edges the fibers protruding from the margin, as is shown in fig. 57, *d, d*. This is much as I have illustrated the residual fibers of bone as doing in fig. 21.

The clearness and regularity of the appearance of these fibers of the cementum in my preparations make it a matter of great surprise to me that they have not been before described by writers on dental histology. I can only account for its oversight by the fact that very few studies of the peridental membrane have been made, and these seem to have been only casual, and thus, the connection of the fibers of the two structures have escaped notice. In this way the appearance of fibers in the cementum has been passed as something not understood, or they have been wrongly interpreted. However, most of the studies of this structure have been made from dried sections in which the fibers could not be demonstrated.

CHAPTER XIII.

IRREGULARITIES IN THE GROWTH OF CEMENTUM.

Hypertrophies of the cementum have been under discussion for many years and generally they have been regarded as pathological phenomena. I think, however, that the careful student must admit that, in the vast numbers that occur, there are comparatively few instances in which the pathological character of these is fairly made out. They have been regarded as connected with all manner of aches and pains. I wish now to call attention to a mode of study of these, which, if followed, will, I think, dispel most of these notions. Not that I wish to affirm that in no case a hypertrophy of the cementum may be related to a process of disease, but rather to show that this is not necessarily the case and as a matter of fact is very rarely so. They are to be regarded rather as irregularities than as pathological phenomena.

I have already said that the cementum is to be regarded as *continuously growing* in the sense that its growth continues to old age. It may be found augmenting in thickness in persons seventy years old and the process be perfectly normal. I wish also to further emphasize the fact that the manner of the growth is by interrupted accretion, or in periods of activity and rest. The intervals of inactivity are probably very great sometimes, but the examination of the cementum of any considerable number of persons at thirty years of age, and comparisons with a similar number at fifty years, will show that there has been a pretty regular increase in the thickness and in the number of lamellæ of the cementum. If those of fifty are again compared with those of seventy,

111

a farther increase in thickness and number of lamellæ will be manifest. This growth takes place at irregular intervals of time, which is expressed in this lamellation. The lamellæ are laid the one upon the other successively and the outer ones are of course the last in the order of growth.

When a tooth presents through the gum and has taken its place in the arch, its cementun will generally present but one layer; but if it has been brought into use for a time it is likely to present two or three, one formed contemporaneously with the root and one probably when it was first brought into contact with its antagonist or possibly while it was being protruded after the growth of the root was accomplished. It seems probable also from examinations I have made, that there often several layers of cementum deposited during the movements of the teeth connected with the lengthening of the face which was illustrated in Figs. 52 and 53. At any rate differences in this regard are observed, whatever be their cause. As the tooth grows older new lamellæ are laid down. It must be admitted that the study of these lamellæ has not as yet been sufficient for us to form any definite idea as to their relation to the age of the individual after the first two or three have been laid down. But however this may be, it will be found upon examination that every case of irregularity in growth will be connected with one or more of the lamellæ, and the relative time of the irregularity of growth to the deposit of the individual lamellæ can be made out.

On applying this mode of study to the irregularities of growth in my collection, I find a great variety. They are connected with the lamellæ in all sorts of ways. Some belong to a single lamellæ, others include several, while others again include all of the series from the first to the last, each one being thicker in the hypertrophied portion than elsewhere. Now it is perfectly evident that such a hypertrophy, as this latter, has been forming with each successive growth of the cementum from the first to the last, while those that are confined to one or a few lamellæ have begun and ceased with the deposit of these. There is no such thing as interstitial growth of the cementum, and no thickening of the lamellæ can occur after another is deposited over it.

Among these hypertrophies confined to one or a few lamellæ, the greatest variety will be found. I have specimens from teeth just erupted showing hypertrophy of the first and only lamellæ yet deposited. But these are more rare than those connected with the second or third. It seems to be with the latter that the greater number of the irregularities are connected ; though a goodly number will be found beginning with those deposited later ; and some are connected with the last one, even in very old persons. These last have of course occurred late in life while the others have occurred at an earlier age. The greater number of the irregularities in the deposit of the cementum are probably connected with some especial strain upon the tooth, and their causation probably corresponds with that of the absorptions which are yet to be studied. We generally find these combined in the same tooth and occurring at about the same time. That is to say, the absorptions of the cementum are shown in the lamellæ that lie next beneath those that show the condition of hypertrophy but are generally upon another portion of the root, which may be contiguous or upon the other side. I frequently see these latter that have cut away the first two lamellæ, penetrated the dentine to some depth and have been refilled with cementum in connection with the deposit of the third or fourth lamella. Now these facts have prompted this thought; the tooth makes its growth and presents its crown to its antagonist. At first the cusps of the one do not strike fairly into the sulci of the other. This causes a lateral strain upon the

15

peridental membrane as the tooth is forced to one side
sufficiently for the proper adjustment of the cusps. A
portion of the membrane is put upon the stretch and
probably the cementoblasts are stimulated to increased
deposit of cementum during this interval. This results
in an irregularity of growth which may be in connection
with a single general lamella of cementum, or it may be
only a partial lamella confined to one part of the root.
At the same time absorption may have occurred in another part as upon the opposite side, removing some portion of the layers previously formed, or forming irregular
openings throughout the whole thickness of the previous
formation and penetrating the substance of the dentine.
These latter are now refilled with cementum with the next
lamella deposited, and afterward the deposit may take
place regularly over both the hypertrophy and the absorption area, and these will be found covered with a number
of regularly formed lamellæ.

Such a theory seems very weak, however, when confronted with thickenings like the one shown in fig. 60,
which is confined to the first layer, while the subsequent
ones are very regularly formed. This was a thickened
portion on the side of a root of a cuspid tooth. It can not
be stated positively that this was all formed before there
was a contact with its antagonist, but its deposit was
certainly continuous with the layer formed contemporaneously with the development of the tooth. I have
found such thickenings of the first layer in teeth not yet
fully developed upon which there was as yet but the one
layer of cementum deposited. This seems to show that
these irregularities do not depend wholly upon extraneous
influences.

Other instances will be found in which on some part
of the root there will be a thickened portion confined to a
single layer belonging to a much later date, as the one
illustrated in fig. 59. This was on the distal side of the

root of a molar near the gingival border, and so far as could be made out without a history of the case, the conditions point to an irregular strain upon the tooth as a cause. The crown of the adjacent bicuspid had been broken away a number of years previously, apparently, and this tooth had leaned forward over the remaining root. Upon the mesial side of the anterior root there were several absorptions affecting the lamellæ next beneath the one hypertrophied upon the distal side. Appearances indicate that this occurred at an age of upwards of fifty years.

Other cases occur in which there is an increased thickness in each successive layer over the same spot. These are found mostly about the apex of the root upon one side, or covering the entire apex in the form of a rounded knob. They may become thinner gradually toward the neck of the tooth or cease abruptly. In the latter case absorption areas will generally be found around the border of the hypertrophied portion. I have illustrated such a case in fig. 61, from a hypertrophy of the root of a superior bicuspid. It will be noted that in this case the roots were originally separate, and that they became fused together with the deposit of the second lamella of the cementum; and that this lamella presents the greatest thickness, while those deposited later are progressively thinner; therefore a large part of this deposit must have occurred early in life. Connected with this, areas of absorption have occurred at *d, d, d,* which have narrowed the root at that part, increasing the nobbed appearance of the apex. It will be seen that these absorption areas have been refilled with cementum in connection with the deposit of certain lamellæ.

I have seen a number of cases of hypertrophy similar to the one last described, that I supposed resulted from the loss of an antagonist and a consequent partial protrusion of the tooth from its socket. But most of these

showed, when sections were made, that the thickening
had begun very early in the history of the tooth. The
second and third lamellæ being the thicker, necessarily
excluded the loss of the antagonist as the cause. Other
cases occur, in which the increased growth belongs to the
lamellæ last deposited, and in these the loss of the antagonist would seem to be a probable cause. Our knowledge
of the subject is not yet sufficient for the satisfactory
determination of the cause of these irregularities in any
case, but the suppositions given may lead to further study
and thought, and lead to something more satisfactory.
In the meantime it seems evident that the anomaly should
not be considered as a pathological state, but rather as an
irregularity of development. Yet these enlargements,
when considerable, may impinge upon the surrounding
tissue in such manner as to induce conditions of a pathological nature.

Another point of interest should be noted in this connection. It is a well-known fact that cementum and
bone never unite. At least no well authenticated case is
on record. In reptiles and fishes the osseous union of the
teeth with the bones is the normal condition. With this
fact before us, and considering the great similarity of
cementum and bone, it seems quite remarkable that such
a union should not sometimes occur. In the study of this
subject I often find an exceedingly thin peridental membrane dividing the hypertrophied cementum from its alveolus. But there is always some soft tissue. I do not
remember of any case occurring high up on the root
about which the membrane was unusually thick. The
rule is that it is thinner, than about the parts not hypertrophied. The bone forming the alveolus is also apt to be
more than usually cancellated.

When cementum in its growth approaches cementum
the case is entirely different. On coming together fusion
occurs whether the roots belong to one tooth or to differ-

ent teeth. In this way the roots of neighboring teeth become fused together in a considerable number of instances. Sometimes this seems to have occurred contemporaneously with the development of the teeth, and the subsequent thickening of the cementum obliterates the point of junction to such an extent as to give the appearance of a single root with two crowns. But many of the cases seen have evidently occurred comparatively late in life. I have seen a number of specimens in which it seemed that the fusion of the roots had occurred on account of the teeth having been forced out of position, especially in molars, which had inclined forward after the loss of the next tooth anterior to them, and the roots pressed backward in such a way as to come in contact with the roots of a posterior tooth. Again spherules of calcific material occur in the peridental membrane. As has been noted, cementum may be deposited upon these, and this will fuse with the cementum of the root of the tooth.

These facts taken together seem to show that, while cementum and bone are so very similar in structure, there is a radical difference in the specialized cells by which they are formed, which prevents them from coalescing in their functional activities. Yet these cells are developed within the meshes of the same membrane, the fibers of which span the space from the one hard formation to the other.

CHAPTER XIV.

The absorptions occurring in the alveolus are of much interest and practical importance to the practitioner. They are very frequent, occur under various conditions and circumstances, and may be of any extent, from the slightest erosion of the surface of the root of the tooth, or of its alveolar wall, to the complete removal of either or both. It is by absorption that the roots of the temporary or milk teeth are removed to give place to the permanent or adult teeth. And so far as microscopic study of the subject can determine, it is by precisely the same plan that the roots of permanent teeth are occasionally absorbed, either in part or completely. So far as has yet been determined, it is this same process of absorption that is the great enemy with which we have to contend, in the various operations of replanting, transplanting and implanting natural teeth.

The subject of the absorption of the roots of the temporary teeth does not properly come within the scope of this work, except incidentally for comparison with other absorptions. A study of the physiological errors that occur in the absorption of the roots of the temporary teeth will do much to explain some things that seem very strange in the absorptions which occur in the alveoli of the adult teeth. By physiological errors, I mean, as has previously been said, an action of the tissues which is purely physiological in form, but going beyond the needs of the time and perhaps calling for a counter-action on the part of the adjacent tissues for its correction; but not going to an extent that can properly be classed as pathological.

118

In the consideration of the soft tissues such errors are not readily detected, because all traces of them is soon effaced; but in the study of the bones, and especially of the teeth, where these errors remain written indelibly in the structure in which they have occurred, possibly many years before, we may, after sufficient observation, trace their progress and subsequent correction with almost the same certainty that we can trace a wellworn pathway through the wooded hills.

The process of absorption has been spoken of, its peculiar cells illustrated and described, and its effects upon the hard tissues detailed, in the previous pages. It is generally performed, we may say always, when any considerable mass of hard tissue is to be slowly removed, by the specialized cells known as osteoclasts. How these cells perform this function is not yet perfectly clear. It seems that they elaborate and evolve from themselves a substance which dissolves the hard tissues with which they may be in contact. Some observers, as Krause, regard this substance as being lactic acid, while many others seem unwilling to express an opinion. The action upon the hard tissue is certainly different from that of lactic acid, in that there is very little softening of tissue upon which it acts farther than the portion actually liquefied and removed. The surface, being absorbed, is thrown into elevations and depressions by the form of the cells acting upon it, but the surface of each of these depressions will be found to be clean, smooth and hard, and when dried will glisten like a polished surface. The osteoclasts are not attached to the surface of the bone or tooth by any mechanical means whatever; they simply lie against the surface and are detached with the least movement. They act, however, only when lying in contact with the surface. Any intervening substance whatever will prevent their action. Therefore, the formation of an abscess with pus lying about the end of the root of a temporary tooth, so long as it lasts, is a bar to the absorptive process. This may act in two ways. 1st. The presence of the pus may separate the cells from the tooth's root. 2d. The pathological condition may prevent the physiological action of the cells. The process of absorption is always to be regarded as physiological, but the error of direction and extent may be so great as to constitute a pathological condition, as when the root of a permanent tooth is wholly, or in a great part, removed by this process.

The tissue acted upon in absorption is always passive. On this point there seems to have been a difference of opinion, some writers supposing that the absorption of bone was performed in part by the bone corpuscles. There may be some forms of disease of the bones in which this is the case, as claimed by Cornil and Ranvier; but certainly there is no such thing in the physiological absorptions. The root of a tooth that has lost its pulp, and consequently the vitality of its dentine, will be absorbed as readily and as completely as the living tissue provided always that the tissues in contact with the root be in a physiological condition. I have frequently noted the absorption of the root of a temporary tooth after the healing of alveolar abscess; but, if the abscess continues, the absorption will generally fail in part or entirely. For the performance of absorption, then, it is required that the physiological action of the cells be not seriously impaired. At the same time, the clinical history of cases seems to show that a moderate degree of irritation or inflammatory action may hasten, or even be the condition of the beginning of many of the absorptions. It is not yet clearly made out that the absorption of bone in conditions of inflammation is always the same process as that which occurs physiologically, but I will say that in all the absorptions within the alveolus which I have yet examined, the process has been identical with the physiological

removal of the roots of the temporary teeth; but is mani- -
fested in directions and in forms that are often erratic in
the extreme.

I have examined very closely the condition of the bone
corpuscles in the immediate neighborhood of areas of
absorption in various regions and conditions, and have
never seen any evidence whatever that they took part in
the process. I have, in a number of instances, found the
bone corpuscle uncovered by absorption and their processes removed up to the body of the cell, and yet no
change could be discovered in the condition of the cell
itself. There is certainly no enlargement of the chamber
in which it lies. It seems to be entirely passive. Precisely the same is true of the cement corpuscles. The
dentine is also removed, and the dental fibres cut away
without the least change occurring, either in the remaining parts of the fibrils, or the dentine of the neighborhood.

All of this illustrates the fact that these tissues when
once formed become passive agents. Not that all of their
parts have become inert material devoid of life, but they
are for the most part composed of formed material, in
which physiological activity is reduced to a minimum.

Many of the irregular phases of the absorptive process
might be illustrated by the examination of the roots of
the temporary teeth. Perhaps very few of these are
regularly removed proceeding from the apex of the root
to the crown. Indeed, the more common form is for the
absorption to begin some distance down on the side of
the root, cutting a deep cleft. Then it will begin at
some other point and do the same thing, and at another,
and so on. These will finally merge into the great gap in
the substance of the root, and the process will perhaps
proceed more rapidly in the destruction of the remaining
portion. We have no evidence that during this time the absorptive process produces any inconvenience to the young
animal, or the child. It is not until the tooth is materially

loosened by the loss of its root that inconvenience is felt. But during the earlier part of this process it seems to proceed with much uncertainty and indecision (if such terms are admissible), for we find many instances in which the absorption has proceeded for a time and then ceased—not only ceased, but the work of repairing the breach has been undertaken by the building in of new cementum.

I offer an illustration of such a case, taken from the temporary tooth of a pig, in fig. 62. In this case a large breach extending far into the dentine had been made in the side of the root, nearly midway its length, by absorption, and at *f* the bone had grown forward toward the absorbed area. Now a change occurred. Cementum is again deposited for the repair of the breach, and this is laid down over the cut ends of the dentinal canals, upon the dentine, covering it over smoothly and evenly in this case, though it is not always done so regularly. It will be noted that the gap in the dentine is not repaired by a new formation of dentine. Such gaps are always repaired by cementum, if repaired at all. In many cases there is a much greater deposit of cementum of repair than in this, but this one is sufficient to show that cementum may be laid down upon the dentine denuded of its cementum, which is a point of no mean importance in these days of the study of the various forms of replantation, and of the amputation of roots of teeth. Fig. 65 illustrates the same thing as occurring upon the root of a permanent molar.

Passing now to the permanent teeth, I will first notice the absorptions occurring in the alveolar wall. These are very numerous, and may be studied by preparing sections of any of the teeth of the adult; but the best studies will be had from the alveoli of teeth that are at the time undergoing change of position from any cause, such as the loss of a neighboring tooth, continued pressure, or the incisors (and I suppose the molars also) during that change

of position which occurs during the lengthening of the
face, which was illustrated and described in chapter XI.
Under any of these circumstances changes in the alveolus
and the attachment in the principal fibers of the peridental membrane occur, and these seem to call for absorption and rebuilding of bone. In fig. 63 I present an
illustration of this, taken from the middle portion of the
anterior wall of an incisor. The upper portion of the illustration is toward the crown of the tooth. This illustration shows especially well the method by which the
fibers of the peridental membrane become detached and
reattached during movements of the tooth in its alveolus.
No very considerable absorption areas are seen, but groups
of osteoclasts appear at very frequent intervals, as shown
at d, d, d, which lie in the lacunæ of Howship, absorbed
into the surface of the bone. At all such points the fibers
are detached. Indeed, these fibers seem to disappear with
the appearance of the osteoclasts, but wherever the bone
is not covered by these cells the fibers are found to be in
position. At f it will be noted that a portion of new bone
has been built on to the old, in which the ends of the
fibers are secured. In this way, it seems, absorptions
and changes in the alveolus may occur slowly, or even
with considerable rapidity, and sufficient attachment of
the principal fibers of the membrane be maintained to
hold the tooth securely while its position is being changed.
Parts of the fibers are cut away and some portions of the
bone removed, then the fibers are reformed and built into
the wall of the alveolus by a new deposit of bone about
their ends. These changes are not confined to young
animals, or young persons, but may be found in progress
in the old as well, but are generally more irregular. I
have not had the opportunity of examining a case in
which the artificial movement of the teeth, as in the correction of irregularities, has been made, but from what I
have seen I suppose that the absorption and rebuilding

occurs in precisely the same way. However, in the rapid movements that are often made in these cases, there must be a solid line of absorption along one portion of the alveolus (that pressed against) detaching the fibers *en masse*, while the fibers on the other side are lengthened. Hence, the tendency of the tooth to return to its old position until time enough has elapsed for a sufficient reformation of its alveolus and the reattachment of its fibers.

In adults evidences of changes in the alveolar wall may be found about almost any tooth (so far as my observation has extended) that has changed position from the loss of neighboring teeth. In fig. 64 I present an illustration from the alveolar wall at the posterior surface of a bicuspid that had moved backward slightly from the loss of the crown of the second bicuspid. The bone, *b, b*, seems to have been built in to supply an area of absorption that was considerably more than the needs of the actual movement of the root. That this has been an absorption is clearly shown by the Haversian systems of the bone being cut into and portions of their rings removed, as is shown all along the line. At *e*, a recent absorption has occurred, and from the presence of three osteoclasts (⁺) it is seen to have been in actual progress at the time of the death of the individual. Such absorptions as this latter are not infrequent in the alveolar walls. They seem to occur without any cause that I have been able to trace, though it is probable that they are stimulated by some slight movement of the tooth, and have proceeded beyond the needs, and are again refilled by the deposit of bone.

In a large number of examinations very many spaces will be found at which there seems to be no attachment of the membrane to the bone, and yet the appearance of residual fibers within the bone shows plainly that the fibers have previously been attached here. In these cases there is sometimes evidence of absorption of the surface of the bone, sometimes not, but it seems most probable that the fibers have been removed by this process, though this may occur from some process not yet noted. Precisely the same thing occurs along the surface of the cementum, sometimes evidently from absorption of the surface of the cementum, but sometimes such absorptions cannot be demonstrated. Absorptions of the cementum are not so frequent as those of the alveolus.

In fig. 65 I present an illustration exhibiting the evidences of absorption of portions of the root of a lower molar. In absorptions of the cementum in cases in which it has not been so great as to obliterate the lamellæ we may do much in the way of fixing the time of the occurrence relatively to the laying down of the individual lamellæ, in the same way that we can fix the relative time of the formation of hypertrophies of the cementum described in chapter XIII. These absorptions are found to have broken through certain of the lamellæ and extended, perhaps as those shown at d, fig. 65, considerably into the dentine. .They are afterward repaired by the deposit of cementum, and the lamellæ of cementum subsequently laid down are seen to pass over them without material disturbance. In all such cases we know that the absorption has occurred very early in the history of the tooth, otherwise it would have broken through the lamellæ deposited later. In the study of the subject we shall find these beginning with any of the lamellæ of the cementum, from first to last, as the absorption has occurred early or late in life. In fig. 66 a pit-like absorption has extended from the surface through all of the lamellæ of the cementum except the first, almost reaching the dentine. This was from an old man, and was evidently very recent, for the process of repair seems just begun and is apparently in active progress.

The greater number of absorptions that I have studied seem to have begun in the second or third distinct lamellæ,

and have probably been contemporaneous with the first
use of the tooth, at a time when it is forced a little to
this side or that for the filling of its cusps into the sulci
of the opposing tooth. At *e*, in fig. 65, an absorption of
much greater extent is shown. This seems to have cut
away the entire apex of the root. Absorptions as extensive as this are much more rare than those previously
noted, but close observation of teeth extracted will within
a few years reveal a goodly number of such. They are
found in teeth that seem to have rather short, thick
roots, often with an irregular surface. Sometimes these
will be found upon microscopic examination to be recent
absorptions in which the dentine is exposed. Again,
they will be found covered with a fresh deposit of cementum. In the greater number of cases a close study of the
lamellæ of the cementum will give a clue to the time of
the absorption. For this purpose it is necessary that the
section be carried directly at right angles with the
lamellæ, for otherwise they will not appear distinctly.
It is therefore practically impossible to study every part
of a root. But generally enough sections can be had
from lengthwise cuts to give a good idea of it.

In fig. 65 a study of the lamellæ of the cementum
shows that the absorption which shortened the root at *e*
occurred early in the history of the tooth, and that it was
promptly recovered with cementum. The incremental
lines do not appear very plainly in this part, but they
lead into it in such a way as to leave no doubt. In other
cases that I have examined, that were outwardly similar,
and which might be illustrated, the absorptions have
occurred late in the tooth's history, the absorption having
broken through the greater number of the lamellæ, or
have been recent, as the absorption at *f*, fig. 65, in which
there seems to have been no effort at repair.

From the examinations that I have made I am led to
the opinion that absorptions of this nature in the roots of

the permanent teeth do not remain long without the occurrence of the reparative effort, if the tissues are in a condition for this effort to be made. It may also be stated that, if the tissues are in a condition to produce absorption, they will also be in a condition to make the repair, provided no impairment has occurred in the meantime. Fig. 66, from a section cut from the immediate apex of the root of a cuspid, shows something of the extent and completeness of these repairs.

A class of absorptions precisely similar to that illustrated in fig. 66, is of rather frequent occurrence near the gingival margin of the cementum. I have called attention to these heretofore, and at various times, in the consideration of caries of the teeth, and especially in the appendix to " Formation of Poisons by Micro-organisms," page 168, and in the " American System of Dentistry," vol. I, p. 777. These absorptions are very generally of the form of that illustrated in fig. 66, and when they occur very close to the attachment of the membrane at the gingival border, are liable to become uncovered by the shrinkage of the soft tissues and afford lodgement for micro-organisms, and thus are a predisposing cause of caries. I have often noted quite broad absorption areas at this point which seem to remove that portion of the cementum which laps upon the enamel, producing a marked groove. This is occasionally more extended, cutting considerably into the dentine, and in case it becomes exposed, gives the opportunity for the girdling of the tooth, in whole or in part, by caries becoming implanted in it. I have seen several instances in which the tooth was almost severed from its root by these cervical absorptions. One lower molar in my possession had an absorption beginning upon the mesial surface, that invaded the pulp cavity. In another case now under observation such an absorption so weakened a lower incisor that the crown broke away. A number of similar cases might be mentioned.   These might be mistaken for caries, if the condition of the surface, and the tissues filling the space, were not carefully observed.   But the condition is so different in the two cases that a mistake should not occur.   Of course, after caries has once invaded the part, there is no means of knowing whether an absorption began the breach or not.

From the studies previously cited it seems that the detachment and reattachment of the peridental membrane in parts here and there is continually occurring. Not only is this the case where there has been appreciable absorption of the cementum, as in the cases illustrated, but a study of the fibers included in the cementum shows unmistakably that they have been broken at many of the incremental lines when absorption cannot be demonstrated.   In these cases the constant reappearance of the fibers in the lamellæ subsequently deposited shows that the plan of the reattachment is by new deposits of cementum upon the old.   In this new deposit the ends of the fibers are imbedded, making a firm hold.   This occurs equally well if the new deposit be upon the denuded dentine, as when it is upon the cementum.   This being the constant method in this class of cases, I must now suppose that in the various modes of planting natural teeth, the manner of attachment to the root is the same. That is to say. the attachment of the tooth depends upon the production of a lamella of cementum covering the root.   This lamella of cementum is laid down upon the root by the tissues in contact with it.   It does not seem to depend upon the vitality of the cementum upon which it is deposited.   It does not grow from the cementum, but from the soft tissues—from the cemento-blasts.   If this lamella of cementum is once perfectly formed, there would appear to be no reason why it should not endure, but the apparent difficulty is to obtain that perfect lamella of cementum ; and the absorptions continue little by little, proceeding from the many imperfect points, until the root is destroyed. This appears from studies now made. Future investigations may reveal new factors not yet noted.

I have now finished the task I set myself to perform —a practical histological study of the periosteum and peridental membrane. The task has been difficult in many respects, and has required an amount of labor much greater than was expected in the beginning. Although I had much available material, I have thought it best to make new preparations of all the tissues. These have all been gathered, and the work done since the first of October of last year. As the work progressed it was found that a number of series of sections were required for study in special directions, which greatly increased the labor. All of the illustrations are made from freshly prepared material. The work is now before the profession, and by the profession its value must be judged. Many phases of the subject are new. Very few studies of these tissues had been made by previous observers. Therefore, extended references do not seem to be called for. Indeed, the literature does not furnish them. Hypertrophies and absorptions of the cementum have been studied by John Tomes, Chas. Tomes, C. Wedl, Salter and others, and among these some very brief examinations of the peridental membrane appear. Among the works on general histology there is some brief mention of the characters of the periosteum and peridental membrane, and several have mentioned the presence of residual fibers (fibers of Sharpey) in the cementum. Such notices have, however, been too sparse to give much information on the subject. However, after studying these papers, the reader will do well to review any and all of these that may be within his reach.

INDEX.

A.

B.

O.

P.